Introduction to Sensory Evaluation of Food Products

INTRODUCTION TO SENSORY EVALUATION OF FOOD PRODUCTS

Shiv Kumar and Noopur Gautam

Delve
PUBLISHING
www.delvepublishing.com

Introduction to Sensory Evaluation of Food Products

Shiv Kumar and Noopur Gautam

Delve Publishing
224 Shoreacres Road
Burlington, ON L7L 2H2
Canada
www.delvepublishing.com
Email: orders@arclereducation.com

ISBN: 978-1-77469-478-7 (Hardcover)

ABOUT THE AUTHORS

Prof. Shiv Kumar is Mtech. and PhD in Food Technology from GB Pant University and Budelkhand University. He is presently working as Head of Department on the position of Professor for the past 10 years. He has 26 years of experience in teaching and research. He has holded many Adminstrative and Academic positions in Bundelkhand Unievsrity as Registrar, Finance officer, Dean Science, Dean Engineering and Pro-Vice Chancellor of the University. He has authored 4 books and 15 research papers in various National and International Journals, many research articles and reviewed research papers for reputed International Journals. He was also awarded best teacher by Governor of UP, India in 2001. Recently, he is coordinating and handing many research projects in Food Technology.

Dr. Noopur Gautam has done B.Sc in ZBC from Kurushetra University, M.Sc Food Technology fromGuru Jambeshwar University, Hisar and PhD in Food Technology from Bundelkhand University. She is presently working as Assistant Professor in Institute of Food Technology, Bundelkhand University, Jhansi (UP). She has wide experience in the field of Food Inndustry having worked with Food quality Lab and multinatinational company for more than five years. She is a trained auditor in Quality Management Systems especially HACCP and ISO. Dr. Noopur Guatam is guiding M.Sc and PhD students and has many research publications to her credit. She is also on Elsevier's panel as an expert refree with outstanding contributions in reviewing research papers in Food Technology area. There is an amalgam of research and its application commercially in her approach which reflects in her work.

TABLE OF CONTENTS

List of Figures .... *xi*
List of Abbreviations .... *xv*
Abstract .... *xvii*
Preface .... *xix*

Chapter 1 Customers' Decision Making in the Food Industry 1
1.1. Introduction 2
1.2. Types of Decision-Making Processes 3
1.3. The Initial Purchases Made 4
1.4. Repeat Purchases 7
1.5. Consumers' Purchasing Decision Towards Food Products 9
1.6. Nominal Decision-Making 10
1.7. Limited Decision-Making 11
1.8. Extended Decision-Making 12
1.9. Decision Making on Food Survey 14
1.10. Food Decision Making 16
1.11. Identification of Factors Influencing Food Consumption Decisions .. 20
1.12. Economic Considerations Associated with Food Choice Behavior ... 21
1.13. Impact of Factors Influencing Food Consumption Decisions 22
1.14. The Current Information Environment Influencing Seafood Choices 23

Chapter 2 How Quality Affects Acceptance and Consumption of Food Products 25
2.1. What is Food Quality? 27
2.2. Definition of Food Quality 32
2.3. Taste and Flavor 43
2.4. Appearance 44
2.5. Volumetry 47
2.6. Spectrophotometry 48

Chapter 3 **Challenges in Ensuring Consumer Acceptability and Market Success of Products** **51**

3.1. Functional Challenges in Ensuring Consumer Acceptability and Market Success of Products 54

3.2. Other Functional Challenges a Company Must Confront to Ensure Market Success of a Product 58

3.3. Psychological Challenges 59

3.4. Economic Challenges 63

3.5. Sociological Challenges 68

3.6. Environmental Friendliness 73

3.7. Product Quality and Standards Take Center Stage in Consumer Choice 75

Chapter 4 **The Human Senses** **79**

4.1. Introduction 80

4.2. Human Sensory Organs 83

4.3. Human Sensory Receptors 85

4.4. Sense Organs Threshold 87

4.5. Human Sensory Neuroscience 89

4.6. The Sense of Sight 90

4.7. The Sense of Hearing 93

4.8. The Sense of Touch 95

4.9. The Sense of Taste 97

4.10. The Sense of Smell 99

4.11. Other Human Senses 101

4.12. Challenges Facing Human Senses 103

4.13. Solutions to the Challenges Facing Human Senses 104

Chapter 5 **Olfaction** **107**

5.1. Introduction 108

5.2. The Psychology and Physiology of Athletic Performance in Humans 114

5.3. Theories of Olfaction 119

5.4. Parameters of Olfactory Stimulation 120

5.5. Olfaction-Related Physiological Mechanisms 120

5.6. Olfactory Quality Classification Systems 121

5.7. Innovations in the Sensory Reception of Odor 123

5.8. Investigating Different Conceptions of Radiation 124
5.9. The Shift Theory, Developed By Raman 125
5.10. Theories Regarding Infrared Spectroscopy 126
5.11. Theories on the Pattern of Stimulus 126
5.12. Chemistry Theory 127

Chapter 6 Importance of Sensory Evaluation in Food 129
6.1. Definition 130
6.2. Applications 139
6.3. Advantages 140
6.4. Disadvantages 141
6.5. Sense Analyzes Concerning Dairy Products 142
6.6. The Aroma and Taste of Food Products 142
6.7. Is Sensory Assessment Useful in the Food Industry? 144
6.8. Dissimilarity Exams 144
6.9. The Descriptive Examination is the Second Type of Examination 146
6.10. Emotional Testing is the Final Stage 148
6.11. Morphological Characteristics 149
6.12. Blinds With Multiple Colors 152
6.13. Summary of Sensory Testing 153

Chapter 7 Types of Sensory Evaluation of Food 155
7.1. Introduction 156
7.2. Analytical Sensory Evaluation 157
7.3. Discriminative Tests 162
7.4. Types of Discriminative Test 164

Chapter 8 How Do Off-Flavors Occur? 179
8.1. Introduction to Off-Flavors 180
8.2. The Smell of Chlorophenol May Be Found in Iodine, Plastic, and Vinyl ... 185
8.3. Freshly Mowed Grass With a Strong Aroma 188
8.4. Husky/Grainy 189
8.5. An Introduction to Hygiene 196
8.6. Fermentation's Undesirable Flavors and Their Management 196
8.7. The Most Commonly Observed Emotions 198

8.8. Discussion ... 198
8.9 Summary of Off-Flavors ... 201

Chapter 9 Who Carries Out Sensory Testing? ... 203
9.1. What are the Many Types of Sensory Testing? ... 204
9.2. What are the Sensory Profile Requirements for the Product? ... 206
9.3. How are Sensorial Evaluation Samples Created? ... 208
9.4. Featured Facilities ... 210
9.5. What are the Advantages of Employing a Methodical Approach? ... 211
9.6. Sensory Evaluations By a Trained Panel Vs. Sensory Evaluations By a Consumer ... 214
9.7. The Panelists' Preference ... 216

Chapter 10 Sample Preparation for Sensory Evaluation ... 217
10.1. Effective Sensory Evaluation ... 218
10.2. Understanding How Our Environment Defines Your Tasting Behavior ... 225
10.3. Preparation of the Specimen ... 235
10.4. A Category's Extensibility ... 238
10.5. Descriptive Tests Can Be Used to Evaluate Cookies ... 239

Bibliography ... 241

Index ... 249

LIST OF FIGURES

Figure 1.1: Customers' decision making in the food industry
Figure 1.2: Decision making process
Figure 1.3: Initial purchases
Figure 1.4: Repeat purchases
Figure 1.5: Consumers' Purchasing Decision towards Food Products
Figure 1.6: Food choices
Figure 1.7: Food survey decision making process
Figure 1.8: Food decision making
Figure 1.9: Age and weight status can affect food decision making
Figure 1.10: Individual Influences
Figure 2.1: Image of food
Figure 2.2: Food quality management
Figure 2.3: Scientist measuring food quality of cereals
Figure 2.4: Food preservation
Figure 2.5: Food quality control
Figure 2.6: Food quality research
Figure 2.7: Quality of food
Figure 2.8: Portion control image
Figure 2.9: Food inspection image
Figure 2.10: Food quality control
Figure 2.11: Food labeling according to HACCP
Figure 2.12: Quality management foods
Figure 3.1: Customers are the main influencers and pillars of a business. Their entitlement to the success of a product is key
Figure 3.2: In the effect of product marketing, customers highly determine the success of a product
Figure 3.3: Speed to market is not only a matter on how and when a product is to be delivered but an argument of factors that lead to the delivery of a product
Figure 3.4: The 5 major steps involved in quality assurance

Figure 3.5: In breach of the regulatory compliance, a company faces the threat of losing its customers

Figure 3.6: The various disciplines of corporate sustainability. Corporate sustainability is a challenge in modern day business setting

Figure 3.7: The reluctance in getting products tested turns out ravaging for institutions and business out of the unexpected delivery other than the consumer's needs

Figure 3.8: Poor vocabulary tend to veer away potential customers

Figure 3.9: To understand what the consumer needs is essential in the market success of a product

Figure 3.10: The four key As of marketing that influence consumer acceptability

Figure 3.11: Political boycotts affect the market success of a product

Figure 3.12: The theory behind positive disconfirmation

Figure 3.13: Technology is all about commercialization

Figure 3.14: A healthy environment attracts more consumers and makes them accept the product at their convenience

Figure 4.1: An image showing a brief summary of the five common human senses

Figure 4.2: An image showing the discovery of human senses

Figure 4.3: An image showing the main sensory organs

Figure 4.4: An image showing sensory modality

Figure 4.5: An image the human sensory receptors

Figure 4.6: An image illustrating signal detection theory

Figure 4.7: An image showing an eye which represents the sense of sight

Figure 4.8: An image showing a visually impaired individual

Figure 4.9: An image showing an eye which represents the sense of hearing

Figure 4.10: An image showing a hand which represents the sense of touch

Figure 4.11: An image showing the tongue which represents the sense of taste

Figure 4.12: An image illustrating the sense of smell

Figure 4.13: An image illustrating the sense of tense

Figure 4.14: An image illustrating current technology as a challenge to human senses

Figure 4.15: An image showing the use of contact lenses as a solution to challenges in the sense of sight

Figure 5.1: Olfactory

Figure 5.2: 5+ Olfactory

Figure 5.3: Olfaction

Figure 5.4: The human brain

Figure 5.5: Anatomy of Smell

Figure 5.6: Olfactory Bulb High Resolution
Figure 5.7: Sense Smell
Figure 5.8: Olfactory System
Figure 6.1: Sensory evaluation
Figure 6.2: Food evaluation
Figure 6.3: Sensory analysis
Figure 6.4: Food sensory evaluation
Figure 6.5: Sensory analysis
Figure 6.6: Sensory analysis
Figure 6.7: Sensory analysis of food
Figure 6.8: Food sensory test
Figure 6.9: Sensory evaluation
Figure 6.10: Food sensory image
Figure 6.11: Taste testing
Figure 6.12: Sensory evaluation
Figure 6.13: Taste testing
Figure 6.14: Sensory analysis
Figure 7.1: Types of sensory evaluation of food
Figure 7.2: Analytical sensory evaluation
Figure 7.3: Descriptive sensory evaluation
Figure 7.4: Discriminative tests
Figure 7.5: Paired comparison test
Figure 7.6: The Due–Trio Test
Figure 7.7: The triangle test
Figure 7.8: Rating tests
Figure 7.9: Affective tests
Figure 7.10: The hedonic nine-point scale
Figure 8.1: Flavor Images
Figure 8.2: How to Identify Off-flavors in Beer
Figure 8.3: Different Flavors
Figure 8.4: Flavor matrix
Figure 8.5: Flavor wheel
Figure 8.6: There's No Flavor That's off Limits
Figure 8.7: Bottles of Smirnoff vodka on the shelf in an off license

Figure 9.1: Sensory testing
Figure 9.2: Sensory testing
Figure 9.3: Sensory testing
Figure 9.4: Sensory evaluation
Figure 9.5: Sensory analysis
Figure 9.6: Sensory senses analysis
Figure 9.7: Sensory evaluation
Figure 9.8: Olfaction
Figure 9.9: Sensory senses
Figure 10.1: Sensory evaluation
Figure 10.2: Sensory analysis
Figure 10.3: Sensory laboratory analysis
Figure 10.4: Sample preparation
Figure 10.5: Sensory evaluation
Figure 10.6: Food sensory test
Figure 10.7: Sensory evaluation
Figure 10.8: Sensory integration
Figure 10.9: Sensory evaluation image
Figure 10.10: Sensory organs
Figure 10.11: Sensory evaluation

LIST OF ABBREVIATIONS

CDC	Center for Disease Control and Prevention
CFSAN	Center for Food Safety and Applied Nutrition
CNS	central nervous system
EEG	electroencephalogram
EPA	Environmental Protection Agency
FDA	Food and Drug Administration
HPLC	high-performance liquid chromatography
IFIC	International Food Information Council
IRB	institutional review board
JDS	just recognizable difference
MOB	molecular organization of the brain
PROP	propylthiouracil
QDA)	quantitative descriptive analysis
RDA	recommended daily allowance
SEC	securities and exchange commission
TPC	total plate count

ABSTRACT

Food sensory testing includes the application of human senses in the scientific assessment of food products. Qualities like texture, odor, scent, and taste are examined by trained testers to evaluate product quality or obtain opportunities for development. The test people of a board should satisfy certain prerequisites as food testers: availability, neutral conduct towards the tested product, scientific language competence, no sensitivities or intolerances to test foods, and healthy senses.

This volume presents off-flavor as an unwanted taste, smell, or mouthfeel. This can emerge during storage or handling and frequently causes a decline in the sensory nature of food or drink. Off-flavor happens when dangerous compounds pollute a food because of chemical alterations or adulteration during storage or production or through the development of microbes in a product. For effective sensory testing, sample choosing, preparation, balance, coding, and presentation are crucial to offering an objective assessment. The differences between samples should be negligible; for example, all testers must obtain similar samples for appraisal.

PREFACE

Sensory evaluation is a fundamental part of a food research program or product improvement. The Institute of Food Technologists (IFT) describes sensory evaluation as "A scientific field used to evoke, quantify, assess, and interpret responses to different qualities of food varieties and materials as witnessed through the senses of sight, touch, smell, hearing, and taste." Sensory evaluation tests are often applied in product improvement, quality control, research, and shelf-life research. In every one of these submissions, sensory evaluation information can be utilized to offer direction. A few elements should be considered in performing a sensory evaluation to reduce the chances of error. Sensory tests can be subdivided into three categories based on the data provided. The three varieties are discriminative, affective, and descriptive. The choice of the fitting test must be founded on clearly defined project targets.

Since off-flavors can be caused by an assortment of compounds, with ratios in the ppt or ppb range often adequate to ruin an originally sumptuous taste, they provide a critical analytical test. For consumers, off-flavors can prompt disappointment or even cutbacks, or lapses in purchase behavior. If the product contains some off-flavor, it can be recognized in the sensory analysis laboratory. Sensory evaluation includes the development and application of standards and strategies for reviewing human food responses.

This volume presents it as a 'scientific strategy used to induce, measure, assess and interpret those product responses using the five human senses. The study of sensory evaluation depends on rules for the readiness and serving of tests under controlled conditions so that biasing factors are limited. Sensory evaluation is a quantitative science where scientific information is gathered to determine obvious relations between product qualities and human taste. Data produced from human witnesses are, in many cases, exceptionally variable, and boards of people are regularly a heterogeneous tool for the production of information. To evaluate whether the relationship witnessed between product qualities and sensory reactions is probably genuine and not simply the consequence of imagination, measurements are used to assess evaluation data. Same as other scientific test processes, sensory evaluation requires accuracy, sensitivity, and the prevention of false positive outcomes.

A simple descriptive test seeks to define the sensory criteria with regards to smell, taste, appearance, and texture. The descriptions can be selected by the tester or chosen from a list and must not be biased. As for comparative sensory testing, multiple samples are contrasted against one another. The samples are then rated based on variables like texture, taste, or scent. The evaluation is generally done utilizing the Friedman test, which indicates whether there is a significant difference between different samples. The triangle test is another method whereby three products are provided to a test board. Two products are identical, and one is not the same as the others. The three are then tasted to check whether the scientist can find the slightest difference. The paired comparison test is applied to determine an observable sensory difference or resemblance between a pair of samples in one or multiple characteristics. A difference is made between single and double-sided tests. In one-sided tests, the leader understands the difference between the particular products and recognizes which sample a tester would choose.

CHAPTER 1

CUSTOMERS' DECISION MAKING IN THE FOOD INDUSTRY

CONTENTS

1.1. Introduction....2
1.2. Types of Decision-Making Processes....3
1.3. The Initial Purchases Made....4
1.4. Repeat Purchases....7
1.5. Consumers' Purchasing Decision Towards Food Products....9
1.6. Nominal Decision-Making....10
1.7. Limited Decision-Making....11
1.8. Extended Decision-Making....12
1.9. Decision Making on Food Survey....14
1.10. Food Decision Making....16
1.11. Identification of Factors Influencing Food Consumption Decisions..20
1.12. Economic Considerations Associated with Food Choice Behavior...21
1.13. Impact of Factors Influencing Food Consumption Decisions....22
1.14. The Current Information Environment Influencing Seafood Choices 23

1.1. INTRODUCTION

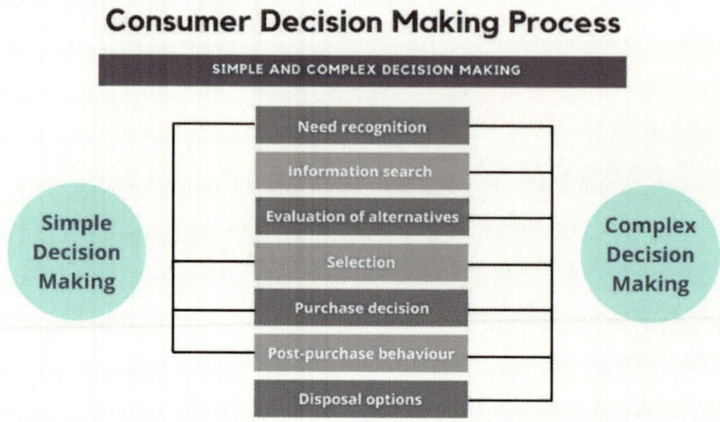

Figure 1.1: Customers' decision making in the food industry.

Source: https://kpu.pressbooks.pub/app/uploads/sites/9/2019/01/Consumer-Decision-Making-Process-2-2-1024x576.png.

Customers can be categorized according to their decision-making styles, which can then assist those customers in making better-informed product choices. The following categories can be used to classify the various methods that can be used to make decisions: It is possible to study consumers using two distinct methodologies: one that emphasizes defining consumer types in general, and the other that has an emphasis on cognitive and affective mental orientations. Both methodologies are viable options. According to Sproles and Kendall, how a person approaches the process of making decisions is comparable to their "fundamental consumer personality," which is comparable to the concept of personality in the field of psychology (Yağcı & Göğüş, 2008). The primary purpose of this study is to categorize customers according to their decision-making style or "shopping personality" as it relates to their food shopping routines. This will be done to better understand their buying patterns.

This study will evaluate the decision-making style of customers, sometimes known as their "shopping personality," to better understand how they buy groceries. The items that customers purchase at the supermarket have a direct bearing on their eating routines as well as the amount of weight that they put on. This is because the decision-making processes of customers play a significant influence on the shopping routines that they engage in. In the same way that marketing strategies can reap benefits from customer segmentation, nutrition education campaigns can reap benefits from customer segmentation that is based on the decision-making style that customers exhibit. In a similar vein, these professionals are now able to direct their efforts toward the education of particular subsets of the population.

Our assessment of the research demonstrates that it has not been used in the study of consumer behavior when it comes to food shopping, even though the consumer decision-making style is widely acknowledged as a substantial influence on consumer purchase behavior. This is the case although the mode of decision-making that a consumer employs is a significant component in the equation of factors that influence the purchase decisions of consumers. The field of general merchandizing is home to the great majority of research done on the many approaches that customers take when making decisions (Ayzeli et al., 2020). Research on how consumers make purchase decisions in stores, research on how consumers make purchasing decisions between cultures, and research on how consumers make purchasing decisions across cultures are all topics that are being discussed here. In this research, representative samples were taken from participants in all three levels of education: high school, undergraduate, and adult.

1.2. TYPES OF DECISION-MAKING PROCESSES

Figure 1.2: Decision making process.

Source: https://www.decisivefacts.nl/structured-decision-making/?lang=en.

The discussion of consumer behavior models stressed several different decision-making processes that customers may utilize when selecting a product or service, with specific reference to the EBM model as an example of one of these decision-making processes. Instead of the EBM model's distinction between extended and limited problem-solving, the Assael model emphasizes inertia, brand loyalty, and limited problem-solving. The Assael model, on the other hand, represented more limited decision-making (Abdulmajid et al., 2014). Additionally, Engel et al. (1968) suggest the concept of habitual decision-making in addition to the three decision-making processes included in the EBM model (Milner & Rosenstreich, 2013). This unique decision-making process is relevant to recurrent purchases. We will evaluate the spectrum of decision-making processes as a continuum, spanning from great engagement to little involvement. In contrast to Engel et al. (1968), who use the word "problem-solving," the researcher elected to use the phrase "decision-making," which is also used by other writers (Milner & Rosenstreich, 2013).

1.3. THE INITIAL PURCHASES MADE

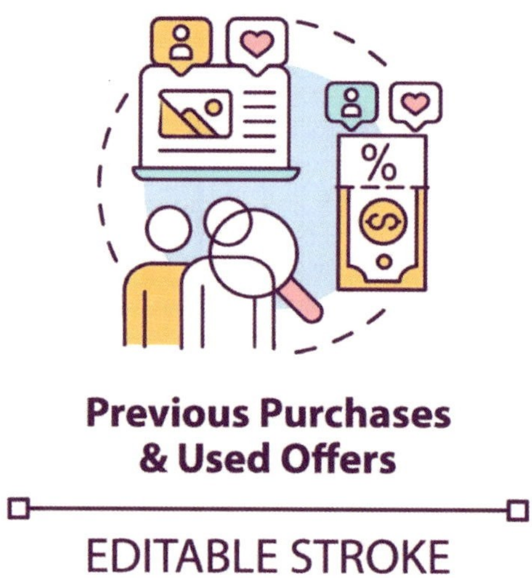

Figure 1.3: Initial purchases.

Source: https://static.vecteezy.com/system/resources/previews/002/254/582/original/previous-purchases-and-used-offers-concept-icon-vector.jpg.

Engel et al. (1968) discovered three separate decision-making processes that occur during early purchases: protracted, moderate, and restricted. Early purchases often result in the establishment of enduring purchasing patterns because early purchases typically require lengthy decision-making processes. Inability to make judgments leads to inertia-based habits, such as buying the same things over and over again rather than trying out new brands. This section will shed light on the differences between the processes by concentrating on the prolonged, intermediate, and constrained decision-making processes. Extended decision making is characterized by a very high level of involvement on the part of the buyer (Yang & Boyle, 2016). A drawn-out process for obtaining a choice includes completing an exhaustive search for information both internally and externally, which is then followed by a sophisticated evaluation technique that assesses several distinct options. The buyer is likely to have second thoughts following the purchase, which calls for an in-depth analysis of the item that was acquired by the customer.

According to Schiffman and Kanuk's research, protracted decision-making happens when clients do not have established criteria for evaluating the product category as a whole or a particular product brand contained inside the product category. As a consequence of this, clients need a substantial deal of information about a variety of things, as well as information that will assist them in building a set of criteria for evaluation. As described by Engel et al. (1968), clients engage in extensive decision-making when they go through a lengthy and thorough decision-making process. When buying a car, a costly item, or a product or service when the cost and perceived threat of making a poor choice are substantial, clients adopt a drawn-out decision-making process. If they're shopping for themselves or someone else, this holds. According to Solomon, the prolonged decision-making process is characterized by a high level of customer risk and engagement, which results in a comprehensive search for information from a range of sources before store visits. According to Hawkins and colleagues, lengthy decision-making may occur without sufficient measurement and evaluation of the required criteria when the buyers' principal concerns are emotional, such as the question of whether or not to go on a vacation boat trip (Warnakulasooriya et al., 2020). It is crucial to highlight that the processes of decision-making are not necessarily conducted chronologically, as reported by Engel et al. (1968) in their study.

Customers may seek extra information regarding where and how to obtain a product in addition to doing an in-depth information search and weighing the advantages of several various possibilities. The following is a summary of how Engel et al. (1968) defines the process of making long-term decisions: "Thought and evaluation precede the act of purchasing and utilizing since the right decision must be made." According to Engel and colleagues, the process of making judgments within a medium falls between the extended and limited decision-making categories. As a result, the amount of information that is essential is quite low, and it is not difficult to acquire. The customer has a wealth of options to weigh before making a selection, thus they must do so thoroughly. The method is usually quick and straightforward to utilize, allowing you to decide a matter of minutes rather than hours or days. A decision that falls somewhere in the center of the road would be somewhat like picking a movie from a menu of alternatives.

Publications like newspapers make it easier for anyone to learn about other possibilities, and a close friend's perspective or an opinion piece written by a critic can considerably affect how someone perceives them. The level of constrained decision-making, according to Engel and colleagues, is the greatest level of the continuum that can be applied to initial purchases. The lack of prior information collecting and evaluation, as well as the consumer's lack of time, resources, or inclination to engage in lengthy decision-making, are all features of this decision-making strategy. Additionally, this decision-making process is characterized by a lack of prior information gathering and review. As a direct consequence of this, the customer will choose to standardize the procedure by narrowing down the number and variety of information sources, options, and criteria for evaluation that are available. According to Schiffman and Kanuk, consumers have already formed evaluation criteria for product categories as well as specific companies operating inside those categories to make restricted judgments (Tsai et al., 2021). This holds for both basic categories of products and the specific firms that come beneath them. Choosing a well-known brand or the item with the lowest price is a popular selection rule when buyers are presented with a restricted number of alternatives. The buyer may decide to go with a different brand as an alternative.

1.4. REPEAT PURCHASES

Figure 1.4: Repeat purchases.

Source: https://lootly-website-assets.s3.amazonaws.com/resource-center/10+Powerful+Ways+to+Increase+Repeat+Purchases-featured.png.

A recurring buy is made regularly and entails making decisions in a manner that is routine or habitual. The term "repeated buy" refers to a transaction of this sort. According to Engel et al, recurrent decision-making happens when a client is required to make recurring decisions over repeated purchases. This is often the product of the customer's discontent with an option that was previously achieved. When a customer's favorite retail location is out of stock or the circumstance changes in another way, recurrent decision-making occurs in addition to what has been explained thus far. So, the customer must now select whether or not to spend the time and effort necessary to discover another solution that satisfies their needs. The second decision-making process connected with frequent purchases is the habitual decision-making process (Tuorila & Monteleone, 2009). Depending on the buyer's initial decision-making process, this decision-making process could take many diverse shapes. It was found in (Engel et al., 1968) both inertia and loyalty to a single brand or company are examples of traits that could impact habitual decision-making. Customers are more likely to follow habitual decision-making than repeated decision-making for repeat purchases because recurring purchases are made based on habits or routines acquired to cope more successfully with life's challenges.

As a result, purchasers are more inclined to follow habitual decision-making than repeated decision-making for repeat purchases. Those customers who have been faithful to a particular product or service for a long time are more inclined to reward the company that sells or provides the service by supplying continued support, according to (Engel et al., 1968). This would be the circumstance if the consumer has been purchasing the same goods or service owing to happiness with the purchase and the service acquired (Sutrisna et al., 2017). According to Belch & Belch, the concept of brand loyalty is an individual's preference for a given brand that results in recurrent purchases of that brand. An example of a brand or corporate loyalty is a client who owns a motor vehicle, is pleased with the service their dealer delivers for routine maintenance, and as a result, continues to supply that dealer with their support. Customers' loyalty to a brand can be difficult to acquire, as any marketer can tell you. Additionally, Hawkins, Best, and Coney (2020) claim that brands whose customers are dedicated to the brand will go to considerable measures to urge their customers to switch brands (Short et al., 2021). It's because consumers have a strong feeling of brand allegiance. Concerning repeat purchases, very little evidence of brand loyalty has been identified, according to (Engel et al., 1968). Most brand loyalty is reserved for goods that can be used interchangeably, like potato chips, which customers enjoy regardless of the brand they purchase. As a result, customers' purchasing behaviors are defined as both inert and malleable. Although there is not much of an incentive to move brands when applying this decision-making procedure, it is possible that this could happen fast if expenses could be cut or if another brand could give something creative.

1.5. CONSUMERS' PURCHASING DECISION TOWARDS FOOD PRODUCTS

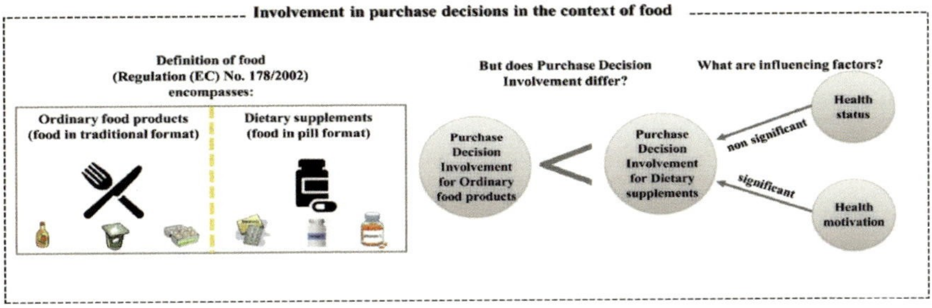

Figure 1.5: Consumers' Purchasing Decision towards Food Products.

Source: https://ars.els-cdn.com/content/image/1-s2.0-S2213434419300374-ga1.jpg.

Marketers must have an awareness of the amount of time and effort that customers invest in the decision-making process so that they may successfully reach their target market and improve their relationships with customers. Many years have been spent conducting research in the hope of gaining a deeper understanding of decision-making and its dual nature as an art and a science. The most important aspect of marketing is, in most circumstances, determining exactly what it is that the client wants and figuring out a way to convince them to choose a certain product. This can be a challenge because customers are notoriously fickle (Schiling & Pham, 2012).

The decision-making that takes place behind the scenes of activities that appear to be simple is heavily influenced by a variety of distinct mental processes. A consumer, for instance, could make a choice based on his feelings "on the spur of the moment," or he could choose to choose between carrying out extensive research and giving it great thought (Shinde et al., 2016). Given all of this information, the fact that there are several distinct processes for making decisions shouldn't come as much of a shock to anyone. Customers' choices typically fall into one of these three broad buckets when making a decision:

- nominal;
- limited; and
- extended.

1.6. NOMINAL DECISION-MAKING

Figure 1.6: Food choices.

*Source: https://miro.medium.com/max/750/1*axW_hwTYGVBRdkyqEAuz5Q.jpeg.*

Things that are sold at lower costs are often treated with a lackluster level of attention in many different contexts. They consist of purchases made consistently, acquisitions of a well-known brand or product, acquisitions that need little to no effort, and searches that are carried out on an occasional basis. If you have been using the same brand of laundry detergent for the past twelve years, it is unlikely that you will spend much time researching the various brands of laundry detergent available on the market today unless there is a problem with the product that you currently use. In other words, you won't recognize it, but you'll find yourself picking up the same bottle over and over again.

This is not the starting point for decision-making at any point in time. You may have spent several weeks looking for a laundry detergent that satisfies your requirements of being cost-effective, having a lovely odor, and making your clothes softer. These are all important factors to consider when selecting a laundry detergent. However, due to your ongoing devotion to the brand that you ultimately select, the effect of your purchase will become less significant as time goes on (Sun et al., 2021). When making decisions on nominal amounts, the single most important thing to keep in mind is that an excessive amount of marketing effort may harm one's clientele. This is the element that one must keep in mind above all others.

If a consumer is already planning to buy a product on impulse or because of their dedication to a brand, they don't need to have repeated contact with a brand in which the company asks what they enjoy about the product or what they would change about the product. In today's climate of intense competition, engaging in excessive marketing is not only detrimental to the financial health of your company; it also has the potential to irritate your clients, which is terrible for the reputation of your brand. When purchasing your goods is a simple decision all that is left for marketers to do is concentrate on preserving the good name of their company and its products. To restate it in a slightly different manner:

- Determine whether or not the item can be bought at the stores where the majority of your customers do their shopping. Check this out!
- Maintain a pricing system that is fair to customers, and if there is a decrease in customer interest, come up with creative new ideas.
- You should work on developing marketing strategies that will make consumers uniquely remember your brand.

1.7. LIMITED DECISION-MAKING

Although limited decision-making requires more involvement from the decision-maker than nominal decision-making does, it does not require a considerable amount of study on the decision-part maker's Products with prices in the intermediate price range, purchases made every few months, and purchases of well-known brand names receive a limited amount of attention from consumers. This attention is limited both in quantity and quality. They require commitment, in addition to possibly conducting some research (Serventi et al., 2022). Customers who make limited judgments not only spend less time deliberating about their purchase but also are less likely to seek online testimonials and reviews of the product or service they are considering purchasing. They can instead evaluate the depth of their product knowledge and base their decisions on the conclusions they draw from the information they have available to them.

You can decide to go to a certain store with the hopes of purchasing a particular brand of juice, but after you get there, you might discover that the store carries an alternative brand that you can buy instead. You have a recollection of having enjoyed this juice in the past, but due to your typical spending habits, it was outside of your price range, so you decided to make a short-term adjustment and buy the alternative juice instead. You have a recollection of having enjoyed this juice in the past. Some people may say that limited decisions are the result of a lack of brand loyalty; however, the truth is that they may be linked to anything from the excitement of trying out a brand-new product to the boredom of sticking to a tried-and-true shopping pattern. Those who say that limited decisions are the result of a lack of brand loyalty are incorrect.

To meet the issue of dealing with restricted decision-making, organizations and marketers need to examine the data they have on their clients to identify the factors that influenced decision-making and devise strategies for overcoming the challenge (Silva et al., 2020). The following are some examples of these factors:

- product excellence;
- possibility of obtaining the product;
- cost; and
- packaging design.

After carefully analyzing these data, companies have the option of modifying their promotional activities and product development to emphasize the advantages provided by the product and reduce the likelihood that customers will select an alternative.

1.8. EXTENDED DECISION-MAKING

In conclusion, when it comes to expensive and infrequently purchased things, more thought and effort are put into the decision-making process. They require a significant amount of involvement on the part of the buyer, frequently involve companies or products that the buyer is unfamiliar with, and call for a great deal of thought and research to maintain buyer confidence. For instance, we don't go out and purchase a large television with a flat-screen very often, so when the time comes to make such an investment, we want to make sure that we are making the best choice possible in terms of the brand and the visual quality of the television that we purchase (Rousta et al., 2021). The client is placed in a more vulnerable position as a result of transactions of a significant magnitude, which necessitates a more in-depth investigation into the circumstances.

Before purchasing a television, it is common practice to seek the advice of one's acquaintances and members of one's own family, to learn as much as one can about the various specifications, and to devote a significant amount of time to the pursuit of searching for product reviews and feedback. Contrast this with the practice of purchasing a television solely based on its price or the fact that it is currently on sale. When it comes to affecting decisions that will affect the company in the long run, the most important thing a brand can do is build a significant presence on the internet: this includes having a website, social media accounts, and a blog (Pu et al., 2020).

Maintain a social presence in which individuals can talk about their experiences with your products and respond positively to individuals who say unfavorable things about them.

- Make it possible for customers to quickly access online feedback by disseminating locations where individuals can review their experiences with your products and give testimonials.
- Ensure that your staff is ready to answer any questions that prospective customers might have, and include in-depth product descriptions on the pages that display the various products that are for sale.
- It is important that you provide individuals with instructions that they can follow as well as additional resources that will assist them in making decisions.

Especially when making decisions with a broader perspective and looking further into the future, the use of digital advice and guided selling can be an extremely helpful tool. At the point of sale, it helps in informing the customer, making it easier for the customer to make an informed purchase decision based on the information provided. It helps to provide easy access to support that simplifies the complexity of the decision by either defining the product, highlighting the dependability of the brand, or minimizing the need to seek out reviews and information. This is helpful because it provides easy access to support that simplifies the decision (Peinado et al., 2014). To put it another way, it is beneficial to provide easy access to advice that simplifies the process of decision-making because it reduces the amount of time spent on it. When clients use digital assistance, they have the impression that they are making "limited decisions" rather than lengthy judgments because they feel less exposed to risk when making these types of decisions.

1.9. DECISION MAKING ON FOOD SURVEY

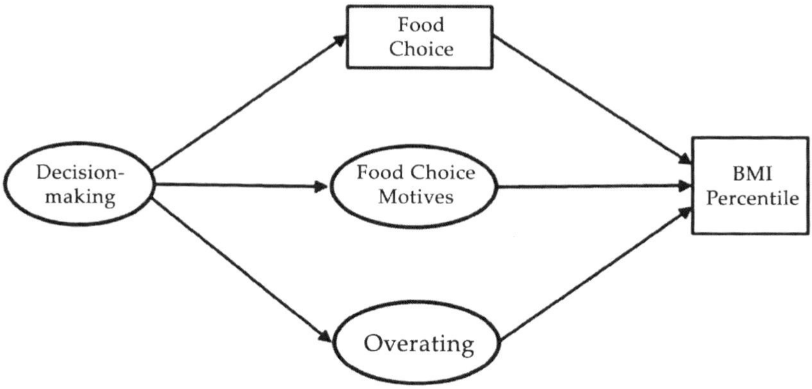

Figure 1.7: Food survey decision making process.

Source: https://ars.els-cdn.com/content/image/1-s2.0-S0195666316307826-gr1.jpg.

The International Food Information Council (IFIC) Foundation conducted its 13th annual "Food & Health Survey," which indicated that "food values" are becoming an increasingly prominent element in consumer purchase decisions, with organic foods gaining favor. The percentage of customers who purchase organic foods and beverages has climbed from 25% in 2017 to 29% presently. When it comes to eating out, the increase is considerably more pronounced: twenty percent of respondents said that they dine at restaurants that advertise organic foods and beverages, which is an increase from fourteen percent the previous year. This is in line with the fact that 37% of customers bought "natural" goods and beverages, up from 31% in 2017, and 26% dined at restaurants that offered natural food and drink options, up from 23% in 2017.

A rise from 50% in 2017 to 59% in 2018 in customers believing that sustainable food production is critical to their purchases and consumption shows that sustainable food production is becoming increasingly important. It's no surprise that the 59% of respondents who believe that sustainability is crucial witnessed significant gains in the two most important sustainability indicators in 2017. In 2018, decreasing the use of pesticides was the top goal for 33% of respondents, which is an increase from the 27% who said the same thing in 2017, while maintaining a cheap food supply climbed to 16% from the 10% who said the same thing in 2017 (Abamecha, 2020).

A person's personal preference and the amount of money they have to spend are still the most important factors. In 2018, the main reasons that drove purchases of food and beverages were, for the most part, untouched from the previous year. An overwhelming 81% of people who took the survey admitted to factoring in the importance of personal preference while making a purchase. This was followed by familiarity, price, healthfulness (61%), convenience (54%), and sustainability. Since 1998, the "Food & Health Survey" has been carried out annually with a whopping 54%. In the modern communication setting, customers are fast to reject phony components, particularly when alternatives are available. When given the choice between two identical products, one of which contained artificial ingredients and the other of which did not, seven out of ten people, or 69%, selected the product that did not contain any artificial ingredients, while one-third of people, or 32%, selected the product that contained artificial ingredients. Participants who chose an organic product were also asked how much more they would be ready to spend for an organic product priced at $1.00 as opposed to one with artificial ingredients, according to IFIC. Some 62% indicated they would pay up to 10% more ($1.10), 42% said they would pay up to 50% more ($1.50), and 22% said they would pay twice as much ($2) if the price went up by 10%.

The situation in which a product is consumed has an effect on consumer perceptions of the relative healthiness of two goods with the same nutritional content. Forty-one percent of those polled thought the "non-GMO" product was better for them, while fifteen percent said the one with genetically modified components was better, and 33% felt a product with fewer ingredients was better than one with more (15%). Both this year and last, the International Food Information Council (IFIC) questioned consumers and asked them to assess the reliability of the many sources from which they typically get information about what foods to eat and what foods to avoid. Among the 14 sources that were identified, the growth of both indicators was driven almost entirely by contributions from governmental organizations. It is estimated that over two-thirds of customers (up from 11% in 2017) regularly get this type of data and 38% (up from 25%) place their faith in government organizations as a reliable source of information. 25% of customers said they trusted government organizations as a resource for information in 2017 (Perera et al., 2014). While last year's question focused on government agencies in general, this year's question asks about specific agencies like the United States Department of Agriculture, the

Environmental Protection Agency (EPA), the Food and Drug Administration (FDA), and the Centre for Disease Control and Prevention (CDC).

Patients have a higher level of trust in medical professionals (66%) than they do in friends and family members (30%). About two-thirds. According to Alexandra Lewin-Zwerdling, vice president of research and partnerships at the International Food Information Council Foundation (IFIC), "Consumers continue to rely on nutrition information sources they profess not to believe." This could be ascribed to the public's uncertainty over the nutrition and safety of food, which is something that we have witnessed repeated times. In two situations of a generational information gap, American people aged 65 and older were more disposed (76%) to trust a qualified dietician than those aged 35 and younger (Predieri et al., 2021). In addition, when asked which information source influenced their perspective on food safety issues, 44% of those 65 and older selected news articles or headlines, whereas just 16% of those 18 to 34 selected the same information source.

1.10. FOOD DECISION MAKING

Figure 1.8: Food decision making.

Source: https://i.ytimg.com/vi/e2CS_RmbqLo/maxresdefault.jpg.

As consumers navigating food-rich environments, we are forced to ask ourselves, "Can I eat this?" if not on an hourly basis, then at the very least on a daily level. If not on an hourly basis, then daily. Is it safe for me to ingest this item, or is there a risk? I'm at a loss for words to explain why I ate so much. I'm not sure. The importance of food to consumers' emotional, physical, and psychological well-being has spurred a substantial corpus of research into the inherent tension in human interactions with food and the decision-making process surrounding food. This research has resulted in the creation of a substantial body of knowledge. This relevance, combined with a social emphasis on our increasingly obesogenic environment, has resulted in the establishment of this field of study (O'Mahony, 2018). This type of research employs a variety of theoretical frameworks to acquire a better knowledge of the elements that influence people's dietary choices. This anthology comprises several pieces that study how our social, psychological, and environmental circumstances, as well as other internal and external factors, influence our eating choices.

The contemporary consumption paradox is the topic of why modern consumers struggle with overconsumption despite having higher levels of food and nutrition literacy than at any other point in history and national concern with calories, fat, and BMI. Consumer research on individual food preferences is assisting us in understanding the modern consumption conundrum. The findings in this domain make it possible to replace the old, authoritarian, and ineffective concept of "food as health" with the more modern paradigm of "food as well-being." The first item in this one-of-a-kind collection is Aydinoglu and Krishna's research, which was written by both of them. It describes how stimulus-based contextual cue size labels (such as "small, medium, and large") influence customers' actual intake of popular goods. "Small, medium, and huge," for example. According to the study's authors, the visual bias of the label influences the amount of food consumed, which might result in accidental or naive overconsumption of food.

This is referred described as "guiltless gluttony" at times. They are still looking at overconsumption, but now Van Ittersum and Wansink (2012) are looking into how our contextual perceptual biases influence our food choices. One of the most well-known examples of a perceptual bias is the Delboeuf illusion. This bias has the effect of making an object appear larger when it is surrounded by a sequence of smaller concentric circles. A series of concentric circles with gradually smaller centers is used to make the target circle appear larger. Van Ittersum and Wansink (2012) created a model that applies this psychological bias to serving platters and food consumption. In addition to the size and color of the dishware, the model is based on the diameter ratio of the serving size. The contextual cues and internal motivational drivers of meal decisions are brought together in two further publications included in this collection (Nishinari & Fang, 2018). The first study, conducted by Finkelstein and Fishbach (2010), looked at the impact of requiring healthy eating on people's subsequent food consumption. The individuals in this study had various degrees of motivation to eat healthily.

Their discovery that feeling driven to eat properly increases feelings of hunger and subsequent food consumption lends validity to the desire for food as well-being as a potentially more effective solution to our society's problem of excessive consumption. Irmak, Vallen, and Robinson (2011) extend their study of the motivational influences on food consumption by demonstrating that dieters rely far more heavily on labeling heuristics than nondieters. This is a discovery that Irmak, Vallen, and Robinson (2011) discovered when researching the motivational influences on food consumption. People on a diet are more prone to base their product assessments and consumption decisions on food-related cues such as healthfulness labels and learned connections with such labels. McFerran et al. (2013), research, which explores the social context of food choices and reveals the unanticipated societal influence that human consumption decisions are prone to having, brings this collection to a close (Nwakalor & Obi, 2014).

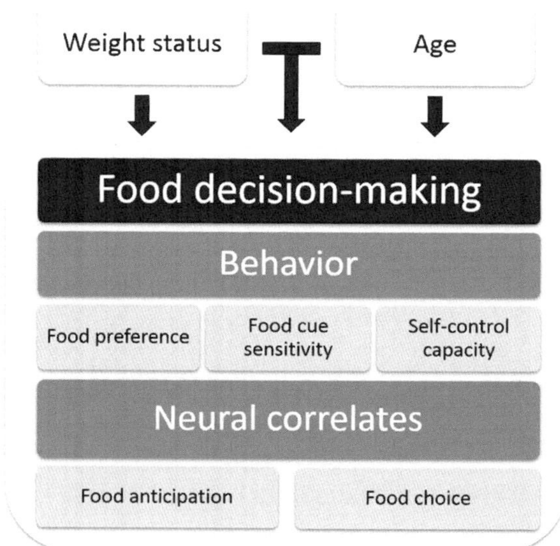

Figure 1.9: Age and weight status can affect food decision making.

Source: https://www.researchgate.net/publication/305719218/figure/fig1/AS:391178261417995@1470275573091/Schematic-overview-of-the-factors-affecting-food-decision-making-which-are-discussed-in.png.

These authors show that once a previous client places a large order, customers tend to order larger pieces; however, they also show that this propensity is influenced by the size of the previous customer. If a former client placed a large order, the likelihood of a new customer placing a larger order is raised. We can ask, "Why did I eat so much of this?" and "How did what other people ate influence the amount of food that I ate?" when we are given a plethora of dietary options and alternatives Size labels on food containers have the potential to significantly influence consumers' mental representations of portion sizes as well as the total amount of food that they consume. Size evaluations are generated by consumers by combining information about the actual size of the stimulus with the semantic signal provided by the size label (Abebe et al., 2006). This enables consumers to form an accurate estimate of the size of the stimuli. Size labels have the potential to affect not only how individuals think about their size and how much food they consume, but also how much food people imagine they consume.

A buyer may be deceived by a product's size label into assuming that one item is larger than another. This is known as "perceived size inversions," and it happens when one thing appears to be larger than another. Customers are more likely to accept a label that says a product is smaller (rather than larger) than the relevant size range. Similarly, buyers are more inclined to accept a label indicating a cheaper price for a product. This is because many buyers choose to buy things that are smaller in size. Customers who overeat as a result of the uneven effect of size labels demonstrate a syndrome known as "guiltless gluttony." Even though customers who use larger plates have proved that they do not serve more food than usual, we do not know what might trigger this phenomenon or who would be most vulnerable if it occurred.

The Delboeuf illusion, according to the findings of five independent studies, may help explain why people tend to overserve when using larger plates and bowls, and a tendency to underserve when using smaller plates and bowls. There is a correlation between these serving biases and the distance between the edge of the food and the edge of the crockery that can be represented as a countercyclical sinusoidal relationship. Changing the color of one's plate or tablecloth is one technique to decrease one's toughness to erase serving bias. Another strategy to lessen one's tough to eradicate serving bias is to improve one's concentration and educational level (Nwakalor & Obi, 2014). The authors open up new theoretical alternatives for connecting illusions to eating behavior by demonstrating that the Delboeuf illusion provides a mechanical explanation for how the size of tableware might influence how much food is supplied and how much food is consumed. These new theoretical possibilities can be viewed as opening new doors. In addition, they investigate the possible impact that tiny changes to the design may have on the health and happiness of customers.

1.11. IDENTIFICATION OF FACTORS INFLUENCING FOOD CONSUMPTION DECISIONS

1.11.1. Individual Influences

Figure 1.10: Individual Influences.

Source: https://s2.studylib.net/store/data/009832908_1-a4c01f-44c2c602d69ff8858487311835–768x994.png.

When it comes to picking what to eat for supper, the overwhelming majority of those polled say that flavor is the most essential thing to consider. Consumer judgments regarding eating or avoiding specific meals, on the other hand, are impacted by a wide range of other human qualities (such as habits, for example). Some people, for example, pick meals not for the pleasure they bring, but rather for the health benefits they provide. The way a meal is cooked, as well as the environment in which it is consumed, can affect both the overall number of nutrients available and the quality of those nutrients (Muhimbula et al., 2011). Other customers' cost, availability, and accessibility issues will probably take precedent over your health concerns. We have no information on the extent to which these factors impact people's meal choices.

1.11.2. Environmental Influences

A person's ability to taste can be impacted in a variety of ways, including their family history and their own life experiences. A range of variables impacts whether or not people consume fish, including cultural norms, the cost of seafood and alternative seafood options, and health and nutrition concerns. People that care about the environment, for example, may choose seafood over other sorts of meat.

A person's food habits are influenced by the evolution of his or her surrounding environment, as well as the person's past. Considerable social and personal events and interactions have a significant effect on the majority of people's choice patterns and life trajectories throughout their lifetimes. The order in which these events occur may have a significant influence on the future decisions made about what to eat. Individuals' responses to both internal and external situations may influence whether or not they choose to make changes to their health and lifestyle (Anton & Luciano, 2007). The complexity of food selection may be examined, for example, from the perspective of a person's life path. As an example, consider the case of a lady who is expecting a kid.

1.12. ECONOMIC CONSIDERATIONS ASSOCIATED WITH FOOD CHOICE BEHAVIOR

There are several ways in which the status of the economy might influence consumers' eating habits. Because of the amount of protein it provides, it appears that seafood is a suitable choice for supper. According to a 1996 poll, consumers in the United States had the lowest income elasticity of demand (the percentage change in demand for a 1% change in income) among 114 nations in the category of "food, drinks, and tobacco." This conclusion was based on the fact that Americans had the lowest demand for "food, drinks, and tobacco." The conclusion that can be derived from this is that changes in income have a relatively little influence on the amount of money spent on food on average. According to Seale and colleagues' research, the United States had the lowest spending elasticity among the 114 nations studied for the category of fish. The percentage change in demand for a one percent change in spending on a category is termed expenditure elasticity (the percentage change in demand for a one percent change in spending on a category). Similarly, the United States had the lowest degree of demand-price elasticities of any of the countries studied (Beekman et al., 2022). The percentage change in demand for a one percent rise in price is characterized as demand-price elasticities (the percentage change in demand for a one percent change in price).

In-depth research conducted in the United States discovered that consumers' adherence to seafood restrictions may be altered if they have extra concerns about their income and price, and the study discovered that this was the case whether or not customers were aware of the regulations. Huang and Lin examined spending and own-price elasticity using data from the National Food Consumption Survey, which was conducted between 1987 and 1988. This was achieved while also taking into account the various socioeconomic backgrounds of the people who consumed the meals under examination. Consumers with low, middle, and high incomes all have different levels of spending and price elasticity. The degree, on the other hand, varies depending on the sort of food purchased. As indicated in the preceding phrase, they feel that a customer's amount of wealth and the quality of the things they purchase are inextricably linked (such as the many cuts of beef available under the category of beef). When we analyze the quantities of money spent on fruits and vegetables by families with high and low incomes, we find that there is a considerable discrepancy between the amounts spent per person and the degree to which demand is income elastic. This is because higher-income households tend to purchase more costly fruits and vegetables than lower-income ones. According to these findings, population averages may disguise significant diversity in how demand responds to changes in income, expenditures, and prices across a wide range of socioeconomic groups (Biró et al., 2020).

1.13. IMPACT OF FACTORS INFLUENCING FOOD CONSUMPTION DECISIONS

Customers will understand and respond to new information in light of their pre-existing ideas, attitudes, and habits. Customers analyzing the health advantages and hazards connected with various types of seafood selections is one example of this. Furthermore, the conditions in which individuals find themselves will impact them just as much, if not more, than the topic under discussion.

There was no association identified between increased fish intake and a reduction in mortality in a long-term, randomized, controlled research trial in which persons with angina were instructed to consume more fish and vegetables. Even with well-designed prospective studies assessing the impacts of delivering advice, evaluating the consequences of offering advice may be challenging since there is a potential that there are factors that mislead the research results. This may make assessing the effectiveness of guidance challenges (Birol et al., 2015). The media can bring public advice to light; but, once it is brought to light, it may be backed or refuted by other policy initiatives, or it may be camouflaged by other news.

According to Ness et al., the improvements in consumption patterns predicted by their research did not always occur. Furthermore, after a choice has been reached, additional measures must be taken to put the modification into action and guarantee that it remains in place. Changing one's behavior, as well as one's food selection behavior (and the resultant impact on one's health), may be explained by a variety of theories. Changing one's meal selection behavior might have a detrimental influence on one's health.

1.14. THE CURRENT INFORMATION ENVIRONMENT INFLUENCING SEAFOOD CHOICES

Because clients now have access to a broader range of communication channels, the decision-making process might be made more difficult for them. Because each of these characteristics is either a store of knowledge or a decision-making facilitator, it has an impact on their ability to make decisions. One of the aspects that distinguish this organization from others is the disorderly method in which information is to be presented to clients. It should come as no surprise that there is a lack of cooperation between public and private organizations, as well as groups with differing perspectives on what constitutes a healthy eating pattern and differing agendas when it comes to delivering advice (Munasinghe et al., 2013). Because multiple recommendations on portion sizes and consumption frequency have been made, it is clear that the federal government has not systematically coordinated consumer counseling on a risk-by-risk or benefit-by-benefit basis. This is proven by the fact that a range of proposals has been made.

Labels and any extra information displayed at the point of purchase in a retail setting are two examples of information environment components that government bodies may choose to control. Menus at restaurants and fast-food establishments are another component of the information environment that may be regulated. The supply of nutritional information and the naming of components Customers may learn about the various components of a meal by reading the ingredient label on pre-packaged fish meals. When referring to specific sorts of items, restrictions may limit the wording that may be used in some cases. While other species of tuna can be called "chunk light tuna," the word "white tuna" can only refer to albacore tuna. All pre-packaged foods sold in the United States are required to contain a Nutrition Facts panel. Furthermore, the display of supplemental health claims and optional nutritional information must follow a set of strict guidelines. There is no need for labels to be applied to perishable items.

The United States Food and Drug Administration (FDA) made guidelines for a voluntary point-of-purchase nutrition information program for fresh vegetables and raw seafood in 1992. Since then, these regulations have been in place. This technique was designed to supply customers with information on the nutritional composition of various items. In 2006, the requirements will be revised to better align with nutrition labeling standards. If a type of seafood is one of the top 20 consumed in the United States, the following nutritional information must be given on the seafood's point-of-purchase label. If seafood is one of the top 20 types, the shop must follow this rule. One serving is three ounces (or 85 grams) of cooked weight and contains no additional oil or spice (Bodyfelt et al., 2008). The proportion of the US Recommended Daily Allowance (RDA) for protein, carbohydrates, total fat, cholesterol, and salt per serving is contained in one serving's worth of cooked weight. The proportion of the US RDA for iron, calcium, and vitamins A and C per serving is also mentioned.

CHAPTER 2

HOW QUALITY AFFECTS ACCEPTANCE AND CONSUMPTION OF FOOD PRODUCTS

CONTENTS

2.1. What is Food Quality? 27

2.2. Definition of Food Quality 32

2.3. Taste and Flavor 43

2.4. Appearance 44

2.5. Volumetry 47

2.6. Spectrophotometry 48

Food quality refers to the aspects of food that are deemed acceptable by humans. This encompasses both outward aspects such as appearance, texture, and flavor, as well as internal standards such as government-mandated grading requirements. In the United States, food safety is governed under the food safety act of 1990. After filing their concerns with trade standards regulators, public analysts get complaint samples as well as samples necessary for normal food market monitoring. Public analyzers conduct scientific examinations on the samples to determine their acceptability. Customers are particularly sensitive to contamination; thus food quality must be maintained throughout the manufacturing process. Many customers need to know exactly what components are in their product for dietary, nutritional, or medical reasons. There are other criteria in addition to the quality of the ingredients and hygiene standards. Customer safety is dependent on keeping all food processing locations as clean as possible. "Meal quality" refers to the qualitative features of meals that are acceptable to (mainly) human clients (Castro et al., 2013). Official grading requirements for eggs, for example, encompass both exterior and inner features, such as size, shape, color, gloss, and consistency (chemical, physical, and microbial).

Figure 2.1: Image of food.

Source: https://media.istockphoto.com/photos/plant-based-vegan-food-for-a-healthy-lifestyle-picture-id1301035781?b=1&k=20&m=1301035781&s=170667a&w=0&h=Mk4zLtRIxQ2QG8w8ObQ8cKUZFSpCLsCIhBNO_H9-YsI=.

Poor sanitation is evident in the most recent e. Coli outbreak linked to spinach in North America, which is still under investigation. Food quality considers the traceability of the food product during a product recall (such as ingredient and packaging suppliers). Labeling also focuses on delivering accurate ingredient and nutritional information (Mouta et al., 2016). A variety of international quality organizations inspect food goods to guarantee that all customers know which products are of the best quality. Before a product can be degusted and authorized, it must first fulfill the institute's selection standards, which include sensory assessment, bacteriological and chemical analysis, nutrition and health claims, and a usage notice. Flavor, health, convenience, labeling, packaging, environmental friendliness, and innovation are all evaluative factors. Because of the reliance on the manufacturing and processing needs of diverse clients, the institute monde selection takes the European Food Law into account. Customers are more likely to accept a product if it is safe to eat and legal to trade in nations with stringent food safety requirements; hence, food safety and legal compliance are also quality factors.

Figure 2.2: Food quality management.

Source: https://media.istockphoto.com/photos/food-quality-control-concept-picture-id1150285143?k=20&m=1150285143&s=612x612&w=0&h=g7fa9p3L9hzszP0FONmE5kFfanGGqzztXtFYUIERRuk=.

2.1. WHAT IS FOOD QUALITY?

"Food quality" is frequently used, although it is difficult to define because a dish's "acceptability" and "value" may change from one client to the next, as well as between cultures and geographical locations. Because quality is subjective if it is not linked to specified criteria or standards, quality attributes or standards are frequently included in product specifications for food items in the food business. The adherence to specific criteria defines quality (Lomarat et al., 2019). Using quality management systems guarantees that food is produced to specifications and that access to competitive markets that need constant quality and a steady supply is maintained.

2.1.1. Quality Management Systems

A company's quality management system (qms) in the food production business lays out the structure, responsibilities, processes, procedures, and resources that go into producing food. Rather than waiting for a problem to happen, a preventive approach uses a continuous process of process and product inspection, auditing, and fault correction to uncover issues before they occur. Three ways are commonly used to guarantee product conformity, respond to customer requests, and assist with due diligence activities: accurate farming techniques (gap) gaps, or good agricultural practices, are fundamental concepts in food safety that aim to reduce biological, chemical, and physical dangers from farm to fork (Lee, 2013). These methods range from site selection to land use to wildlife and habitat conservation to the use of pesticides, nitrogen fertilizers, fungicide fertilizers, genetically modified organisms, and integrated crop management (icm). Superior manufacturing techniques everyone who comes into contact with food, including those who work on surfaces where food may come into contact and those who work with food packaging materials, must adhere to food safety sanitation and hygiene measures. Gmps are required for Haccp, which are normally detailed under applicable law.

The commission regulation (ec) no 2023/2006 defines good manufacturing practices as "those aspects of quality assurance that ensure materials and articles are consistently produced and controlled to ensure compliance with the rules applicable to them and the quality standards appropriate to their intended use by not endangering human health or causing an unacceptable change in the composition of food or causing organoleptic quality degradation." Gmps are the minimal hygiene criteria that must be satisfied to guarantee that a company's items are safe, high-quality, and consistent. The requirements must be followed by everyone in the animal feed supply chain, from farmers to processors to warehouses to distributors (Lee et al., 2021). You must accept responsibility for anything that is under your control. Food outlets, for example, should be constructed to be as safe and hygienic as feasible, with adequate amenities (such as restrooms, drainage, and ventilation) and sufficient plant equipment and facilities (such as pest control and cleaning and sanitation).

Figure 2.3: Scientist measuring food quality of cereals.

Source: https://www.shutterstock.com/video/clip-34780528-food-research-experiment-scientist-analyze-cereal-plant.

Analyze the risk critical control points haccp is a widely-recognized risk-management system that assists food and feed manufacturers in identifying crucial control points for microbiological and chemical contamination, as well as physical and allergy contaminations, such as peanuts and sesame seeds, which may represent a concern. Haccp can identify appropriate control and monitoring mechanisms, particularly for those hazards deemed critical to the product's safety, by evaluating potential hazards in advance and methodically, rather than relying on traditional inspection and quality control methods to determine compliance or failure. In Europe, laws mandate a certain degree of haccp and gmp compliance. It does not, however, describe how the industry should put these criteria into action. As a result, some businesses have created their quality assurance standards, while others rely on government and commercial norms. Food safety concerns of regulators in connection to tbt, the world trade organization (wto) defines "standard" as follows: a document authorized by a recognized authority that specifies optional features or qualities for items, associated procedures, and manufacturing techniques (Chen et al., 2022). Third-party standards give a framework for food firms to develop policies and procedures while also establishing an effective quality management system that fulfills the interests of both consumers and employees while also safeguarding the environment and animals. As customer and consumer expectations, interest in food safety, legal and institutional requirements, and efforts to achieve a competitive advantage have all increased, so has the creation and execution of public and commercial food safety and quality standards worldwide.

A comprehensive standard developed by a customer or a third party will address the following issues: employee training must meet certain requirements. Traceability a high-caliber guide controls for operational routines that must be in place sample and testing timetables system of reporting evaluation procedures it is also critical that a food company's quality management system be proven effective using an approved method. Auditing food businesses are frequently audited to ensure that they are following regulations and internal standards. To ensure that a standard is met, the auditing process includes reviewing documents as well as conducting tests and interviews. Audits performed by second and third parties are all possible. Internal auditors, also known as first-party auditors, conduct reviews within a company. They are used for self-assessment, internal standard verification, and ongoing development. An independent third-party audit conducted by a major firm evaluates the effectiveness of a supplier's quality system in supplying safe materials and products. Refusing to participate in the audit may result in client dissatisfaction. A certification body is a third-party auditor that audits a primary organization and certifies conformity to a specific standard using their expertise and skills. Third-party auditing and certification of the quality management system are used to demonstrate due diligence and/or to market the product or service for which certification is being obtained (Coggins, 2007). As nutritionists began to emphasize the significance of improving food quality, it has become a major focus of their research. The extent to which this is feasible is determined in part by the definition of "food quality." the sum of a food's characteristics and evaluable qualities constitutes its quality. The three most common quality criteria are sensory value, appropriateness value, and health value. Each of the three is performing a subjective assessment. Condition, in addition to the value-related concept of quality, is a value-neutral term.

Figure 2.4: Food preservation.

Source: https://us.123rf.com/450wm/annastills/annastills2006/annastills200600834/150090558-close-up-of-young-woman-in-protective-mask-and-gloves-packing-fresh-vegetables-in-boxes-while-sittin.jpg?ver=6.

Quality is determined by a wide range of variables, making it difficult to pinpoint. Because these characteristics are constantly changing, it is difficult to find definitions that are both mandatory and standard. When it comes to evaluating basic quality characteristics, food producers and processors, merchants and consumers may have vastly different perspectives, which contributes significantly to the existence of contradictions in the food quality issue. Recognizing all legitimate interests, i.e., taking a comprehensive view of all the various components, can assist in resolving existing inconsistencies. A comprehensive evaluation of food quality includes newer and more popular quality categories in addition to the three known partial qualities. Because of customers' impressions, beliefs (prejudices), and expectations, it is often difficult to explain a product's psychological or notional value (Chira et al., 2020). Foods are thought to be selected and consumed based on their properties. Nobody can argue that the cultural significance of food is limited to this specific location. The demographics of those who consume a portion of food, as well as its quantity and cost, all influence its prestige value. Foods that are forbidden or used as a form of reward take on a new meaning in this context. Imports, particularly from developing countries, food production and surplus management, and food aid are all factors that contribute to the political importance of food. Ecological value investigates the environmental consequences of food production and processing, as well as the intricate interconnections and feedback loops that exist between them. Because these additional dietary requirements or characteristics cannot be identified or measured by the product, they can be difficult to find and include. But that doesn't make these food safety regulations any less important. Even as specific food sector interest groups continue to exert influence over quality standards and standards expectations, societal needs and demands are becoming more important. A comprehensive review is being carried out to avoid future mistakes and meet societal expectations for food quality.

2.2. DEFINITION OF FOOD QUALITY

Food quality is a complex and nuanced concept that is influenced by a variety of factors. Quality signals (information stimuli) that influence customer expectations can be internal or external to the product (e.g., the color of fruit indicates ripeness) (e.g., advertising, brand image) (e.g., Advertising, brand image). Higher levels of customer satisfaction imply a greater congruence between what customers expect from a product before purchasing it (anticipated quality) and what they receive (experienced quality) (experienced quality). Positive brand experience has been shown to reduce dissatisfaction and is widely used to infer product quality*. Customers typically perceive brand names as quality indicators because they are well-known (Chowdhury & Das, 2015). According to a survey, buyers place a high value on food flavor and other sensory aspects, as well as health, convenience, and manufacturing technique (organic, animal welfare, etc.). (Organic, animal welfare, and so on.) Sensory components (such as taste, smell, and appearance) were rated as the most important aspect of food quality by 96% of the 30.000 Germans polled, followed by "safety" (93%), and "freshness" (93%). (92%). 92%. In the 2012 'special Eurobarometer 389' poll, a large majority of residents considered food quality, cost, and provenance to be important, but their views on brands were extremely varied.

According to the study, "attitudes toward brand relevance vary more across the member states than opinions on quality, price, and provenance." meals of both high and low quality are possible. Although quality is difficult to define, it is an important factor to consider when purchasing food. The amount and type of ingredients used determine the food's quality. Several member states have expressed concern about similarly labeled and branded commodities on their domestic markets that differ from those supplied nearby. Even if all food information to consumer's regulation regulations is followed, food quality differences may constitute unfair business practices. Customers' food purchasing habits and perceptions of food quality have changed over the last few decades as a result of health and environmental concerns. The goal of this research is to examine the importance of quality, health, and environmental signals to food consumers (Chen & Opara, 2013). We designed and administered a survey to 797 Belgian and Romanian customers to assess their preferences for social, environmental, and gastronomic quality factors. The most frequently cited criteria for evaluating food quality are freshness, flavor, and appearance. Several factors influence how frequently health-related food quality alerts are issued. The packaging, country of origin, and manufacturing process of a product can be used to assess its environmental impact, while its contents, nutrition data, and other additions can be used to assess its healthfulness.

This research shows that Belgian and Romanian consumers can use health and environmental signals to improve their health and the environment. Increasing market pressure (such as requests from more demanding and informed consumers), as well as other factors (such as health and environmental concerns), have both contributed to a surge in interest in high-quality food over the previous several decades. Clients in the twenty-first-century value product quality and health benefits more than ever before. Food quality is a major issue in today's food industry. Assuming that consumer perceptions influence food quality, the current contribution suggests that understanding the cues used by consumers when evaluating food quality is an important step toward mastering the process. Knowing how clients feel about food quality has a huge impact on their purchasing decisions (Lotfy et al., 2021). When customers' perceptions of quality clash with the official definition, for example, or when consumers' interpretations provide a quality impression for food items that do not qualify, the problem becomes even more complicated. Given these assumptions, it's not surprising that food quality ratings differed by time and region.

Figure 2.5: Food quality control.

Source: https://media.istockphoto.com/photos/young-caucasian-serious-supervisor-evaluating-quality-of-food-in-food-picture-id1176094666?k=20&m=1176094666&s=612x612&w=0&h=eLBp4x2ntZnHvMV4fNIiwWyX2_JruRk8gk-TTF1Zt5hc=.

Continuous research is required to capture as much of the current relevance of food quality as possible. Evaluating food quality frequently necessitates two steps. The first occurs before the act of purchasing, while the second occurs following the act of purchase when food is consumed. At the point of purchase, consumers use both explicit (e.g., color, price, and claims) and subtle (e.g., graphic design, material, and color) packaging design signals. According to a European study, information on food labels has a significant impact on consumer perceptions of product quality. Markets require adequate information to function. When there is information asymmetry, it is frequently blamed on the government's inability to enact and enforce relevant legislation, corporate actors' disregard for existing rules, consumers' lack of attention to or understanding of accessible information, and industry marketing strategies (when one side knows more about a product than the other). Labeling is undoubtedly one method of assisting customers in making informed decisions (López-López et al., 2019).

Several studies have concentrated on labeling as a method of communicating aggregated environmental and health product features. It has been demonstrated that increased customer trust and confidence in product information, such as environmental and health warnings, influence consumer behavior. This trust is built by providing clients with more precise, trustworthy, and up-to-date information, such as environmental and health warnings. Customers must evaluate a variety of factors when determining food quality, and they frequently rely on labels that disclose ingredients, expiry dates, health information, and environmental hazards. This study looks at how Belgian and Romanian customers rate meal quality. Furthermore, the consumer preferences survey makes no mention of what "health" means to these customers, even though it is one of the primary motivators for purchasing various goods. Our goal was to determine how much consumers value food features that hint at future health and environmental concerns. Because health and environmental friendliness are not immediately apparent product attributes, they must be achieved through the more palpable interior and exterior indicators.

The study's primary goals are to determine which of the most prominent signals used by consumers to evaluate food quality, healthiness, and environmental impact are most frequently used, as well as to investigate what factors influence how frequently health-related food quality cues are used. This study is especially interested in demonstrating the role of health and the environment in a much larger landscape that consumers use to judge food quality. As a result, customers' perceptions of food quality are rapidly changing as they gain more information about the impact of health and environmental concerns (Kweku Amagloh et al., 2013). Consumer food quality ratings must be transformed into a marketing tool that can be used to influence customer expectations and preferences, as long as a clear goal is in mind. It is obvious that food quality perception is a complicated process, and the use of quality signals is only one of many critical phases, but covering them all is beyond the scope of this investigation. One of the study's most intriguing findings is that it reveals which meal quality metrics customers most commonly use. It also scientifically depicts the indicators people use to assess the health and environmental implications of food, which is useful for marketing sustainability concepts and behaviors.

Figure 2.6: Food quality research.

Source: https://www.newfoodmagazine.com/news/71956/improve-quality-food-research/.

Distinguishing characteristics include those that can indicate the frequency with which food quality signals associated with health are used (ingredient list, fat and salt content, and sugar content) (. Another study attempted to determine which signals were most important in determining the quality of a food product but again failed. In contrast to the current study, their research examined nine quality criteria and divided consumers into two groups based on the quality cues used. Quality signals are important to buyers, even though the current study looked at a smaller number of cues than i did (20 vs. 59). The score was 59–20. A numerical contrast: (20 to 59). (20 to 59). The score was 59–20. The final score was 59–20. Unlike previous studies, this one looked at food quality perception as a whole rather than focusing on one or two specific aspects of the meal. Thinkers' toolkit food qualitative features have been given numerous names in the literature and classified into several classes based on their nature or context. As a result, nelson coined the term "aspects," which distinguishes between the search and experience components of an object (Koli et al., 2022).

Unpackaged tomatoes can be identified by their color, which is a search property, whereas ice cream flavors can only be described by the consumer after they have eaten the product, which is an experienced property credibility qualities are characteristics that cannot be assessed in everyday usage and require special information, such as the environmental benefits of organic apples. The term "qualities" is used to imply that credibility characteristics are defined by how capable a typical consumer is in analyzing a product's genuine performance. The search-experience-credibility (sec) paradigm has dominated consumer behavior research and has been applied to both food and non-food products. Cue utilization hypothesis is an important tool for examining customer perceptions as a paradigm for understanding consumer food quality assessments (Daher et al., 2020). Everything else (such as price, brand, and selling location) is an extrinsic feature that cannot be changed without affecting the product's physical attributes, such as color and form (e.g., price, brand, selling place). Flavor, color, and freshness were found to have a greater influence on customers' overall quality judgments of food items than price, brand, or packaging. In the literature, the term "cue" has a range of connotations.

Use the terms cue attribute an independent attribute. A "natural" product, for example, provides information about the "healthiness" of food, whereas the latter sends information by itself (e.g., being fried or uncooked) (e.g., being fried or raw) Depending on human perception and circumstances, a trait may be classified as a "cue" or an "independent" property. Customers are interested in both what they perceive to be quality signals and what they desire. Because buyers do not use quality indicators before purchase, it is reasonable to conclude that quality attributes are only evaluated after the product has been used. He coined the term "cues" as part of the overall food quality model, and he recognized two applications for it: quality evaluation and food safety (before purchase, when buyers form quality expectations, and after purchase, when customers endure the quality experience) A study that looked at consumer assessments of food quality used the term "cues" similarly. We used Steenkamp's theory and the term "quality signals" in this study because we were interested in detecting food quality at the point of purchase. This study investigated customers' perceptions of their ability to correctly distinguish food quality. Researchers examined not only health-related food quality signals but also how consumers perceive their ability to influence their health through food, which was deemed critical in understanding health-related food quality cues. In the current study, health-oriented food quality was defined as how consumers perceived a food product to benefit their health, and this orientation included functional food elements (Kukovics & Németh, 2014). Because the impact of specific food intake on one's health is a matter of trust, and it is rarely verified after consumption, health-related traits were primarily credibility traits.

Medical, dietary, social, and psychological factors all play a role in a person's overall health. This study did not address food safety or risk, but according to bruns et al., a consumer's health can be divided into two categories: nutritional (e.g., functional foods and lower fat foods) and food safety (e.g., no genetically modified organisms, no gmos). What we do and how we do it many metrics have been implemented in practice to define how customers perceive food quality, and they frequently incorporate both experience and legitimate criteria, such as environmental impact and consumer well-being. Many components were created specifically for this study, according to power et al., while others were drawn from previous research on consumer criteria for judging food quality. Following a literature review and two focus groups, a list of 59 quality signals was developed for inclusion in the questionnaire. The poll included a wide range of quality indicators to gather as much information on customers' food quality assessment habits as possible. A food item's suitability for a specific customer's demands is determined by its packaging and labeling, as well as its sensory characteristics, chemical composition, and physical elements (Kohyama, 2020). Another notable feature is the hierarchical and dynamic interplay of almost all of a food's qualities. As a result, when developing a food product assessment system, the following implications must be considered. According to its definition, the term "quality" can only be considered consistent for a limited time because it is an arbitrary statement, a matter of convention.

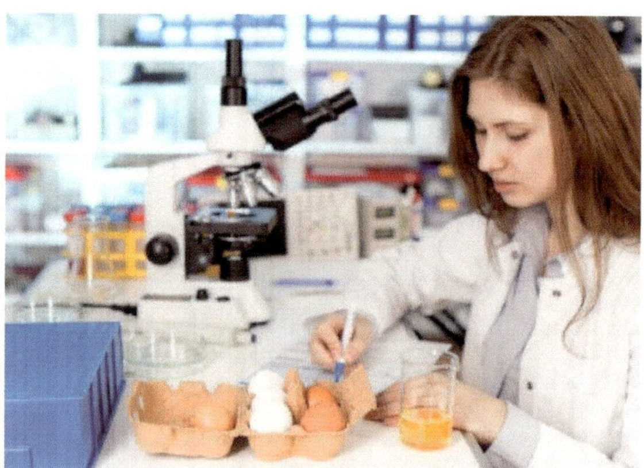

Figure 2.7: Quality of food.

Source https://tecnosolucionescr.net/en/tecno-blog/105-food-quality-the-missing-bit-in-africa-s-1-trillion-food-industry-opportunity.

Customers value the quality and safety of the food they consume. Food quality can be difficult to assess due to its complexity, dynamic change, and subjective nature. The use of systems analysis is critical for problem-solving and making decisions. Various product attributes indicate how well a product's suitability for a specific and stated purpose is represented in its quality concept. The following quality idea is offered to provide a framework for evaluating the food. A food item's sensory characteristics, chemical composition, physical attributes, and the number of microbiological and toxicological pollutants, as well as packaging and labeling, influence its quality following customer preferences. Another distinguishing feature is the hierarchical and dynamic interplay of almost all of the factors influencing food quality. As a result, it is critical to assess the impact of key variables when developing a food product assessment system. We can only declare that quality is consistent for a limited time because it is defined as a convention. It is also affected by the baseline values provided in specifications or standards, as well as the absolute level of product quality (De Kock & Magano, 2020). Only if the baseline values and the conditions that led to their development (including measuring methodologies) are preserved can product quality be altered.

To illustrate our point, we will use the term "etalon," which is a standard scale or space for comparing product qualities. Food quality can be numerically analyzed using assessment techniques such as quality indices. Using these indices as a starting point for developing food assessment systems can help to improve food quality control and product innovation. Scientists must synthesize and communicate information to understand and communicate the chemistry and physical properties of food. Because of the widespread use of high-performance computers for data processing and distribution, previously unimaginable levels of computation have been achieved. A food product's quality can be determined using smell, physical-chemical, and microbiological tests. Traits are visible in some situations but hidden in others. To perform quality assurance and control on food commodities, measurement equipment must also be familiar with these quality criteria.

Factors influencing food quality: Perception aspects sensory qualities and their impact on product acceptance have been extensively researched. These findings can be linked to advancements in sensory evaluation, current understanding of human behavior measurement, and a more comprehensive and professional testing technique. Man eats based on what his senses tell him about the food he is eating. Because these properties are expressed in terms of sensations, they are referred to as qualitative or sensory qualities. Texture, viscosity, consistency, and finger and tongue feel are examples of kinesthetic qualities. This category also includes flavor components or experiences that blend odor and taste. Considerations concerning appearance and shade consumers are more likely to judge the quality and flavor of food based on its appearance, such as color (Kang et al., 2009). The amount of food a person can consume in a meal is determined by the person who is eating it. Color is important in both agricultural products and processed foods. Fruit and vegetable discoloration, both raw and processed, is a common issue. Color changes are frequently accompanied by changes in texture or flavor. Customers perceive off-color foods, such as aged cheese, beer, meat, and fish, to have a lower flavor quality.

Figure 2.8: Portion control image.

Source: https://image.shutterstock.com/image-photo/fresh-salad-measuring-tape-on-260nw-513229240.jpg.

Many fruits and vegetables' color development or variation is closely related to their maturity. A color shift, even if not harmful in some cases, may have an impact on consumer acceptance. Customers may be less likely to buy products if they are dissatisfied with the lack of color. As you can see, many of these assumptions are illogical. Because their natural carotene levels are low, butter, mint-flavored ice cream, and orange sherbet are chemically colored before being treated with artificial green coloring. Artificial colors can be found in oranges, cherries, syrups, and jellies. Colors in food serve several purposes in the human body, including considering everything even if all other characteristics were the same, distinguishing between different foods based solely on color would be extremely difficult (Jain et al., 2014). At least some of these exist. The color of food and where it is displayed may influence people's appetites.

Figure 2.9: Food inspection image.

Source: https://media.istockphoto.com/photos/food-inspector-measuring-temperature-of-food-picture-id930103068?k=20&m=930103068&s=612x612&w=0&h=Tn7jN0i3jlRpb3CAeFwK4tNPyoEt7cg9tzAZ_ApiiVw=.

- Feelings. Beautiful things are desired as delectable, while ugly goods are avoided. The color of food has a strong influence on people's feelings about it. Investing in the future by putting money aside. If you've had similar meals or drinks in the past, you might be able to guess the flavors and characteristics of something new. When we hear about unusual characteristics or unusual meals, we may react differently in some cases. There has been far too little research into the effect of color perception on meal acceptance. Color can be linked to the observer's feelings of acceptance, apathy, or rejection. The effect of colored lights on sensory evaluation has yet to be investigated when color transitions are obscured. This can have both direct and indirect effects, such as changing olfactory, gustatory, or textural perception thresholds

(Difonzo et al., 2019). Several connections have been proposed. The human eye, on the other hand, can distinguish between qualitative and quantitative colors, but it is not a quantitative device.

As a result, color measurement requires specialized equipment. Because of the variety of colors in tomato catsup, mixing it is notoriously difficult. The environment and time of year in which raw materials are obtained can have a significant impact on the final product's color. The texture of food includes characteristics such as roughness, smoothness, and graininess. The viscosity of Newtonian fluids has a direct impact on the mouthfeel of liquid foods. Trained panelists correlated sliminess ratings with the viscosity of gum solution. Fruit and vegetable texture has been studied using devices that measure compression, resistance to penetration, and shear force. The Magness pressure tester is commonly used to determine the ripeness of deciduous fruits, which involves inserting a specific-diameter steel plunger into the fruit's flesh. A variety of penetrometers can be used to test the texture of cooked, canned, and frozen foods (Igarashi et al., 2010). Several studies have found that peach texture influences customer preference, and pectin variations have been linked to such changes. As the fruit ripens, Protopectin converts to water-soluble pectin, making it easier to work with. Protopectin retention in cling and freestone peaches is reduced by shortening the processing time.

Figure 2.10: Food quality control.

Source: https://www.azolifesciences.com/article/What-is-Quality-Control-in-Food-Science.aspx.

2.3. TASTE AND FLAVOR

When something is eaten, the combination of taste, odor, and texture generates a sense known as "flavor." there is no corresponding name in German or many other languages for such broad and full thinking. The flavor is formed of two sorts of molecules: those responsible for taste and those responsible for scents, which are often referred to as aroma and give both experiences. The flavor is the most significant sensory characteristic of many foods. As a feature of meals, drinks, and seasonings, the taste is described as the total of emotions generated by the stimulation of sense endings packed together at the entry of the alimentary and respiratory systems. For sensory analysis, the word for the feelings experienced by the chemical senses from a product in the mouth is limited (Huang et al., 2022). Listed below is a definition of flavor:

- Odor is the olfactory sense of volatile molecules emitted by the posterior napes from a product in the mouth.
- Gustatory sensations (salty, sweet, sour, and bitter) originate from soluble substances in the mouth.
- Chemical sensations (astringency, heat/cooling from spices, bite, metallic flavor, umami taste) that stimulate nerve terminals in the buccal and nasal membranes. Most compounds that produce flavor are nonvolatile at ambient temperature.

They primarily interact with the taste receptors found in the tongue's taste buds. Aroma chemicals are volatile compounds that are detected by the odor receptor sites in the olfactory tissues of the nasal cavity. Scent and taste components should be handled with prudence, since a single component may add to the typical scent or flavor of one dish while creating an off-odor or flavor, or both, in another. A material's olfactory threshold is the concentration at which its odor may still be recognized. Data on threshold concentrations can be employed to compare the strength or intensity of odorous substances. The samples in table 2 illustrate that the olfactory potencies of distinct scent compounds vary by several orders of magnitude. Aromatic compounds' threshold concentrations are dictated by their vapor pressure, which is controlled by temperature and medium values the results are affected by the testing procedure and/or the sensory panel's performance (Hong et al., 2008). Such debates are largely responsible for the constant revisions in threshold values in the scientific literature. Characteristics physical appearance, color, and texture are the most critical physical aspects that determine food quality.

2.4. APPEARANCE

The size and form of food products, as well as flaws and color, have a great impact on first impressions. Sadly, these sensory qualities are generally neglected or discarded. Size and grading also benefit from esthetic components, since they foster consistency and improve the selling process. Grading food by size saves the time needed for further processing steps and increases the quality of the final product. Color is a qualitative feature that has a major impact on the esthetic appeal of a thing. It is connected to numerous favorable and bad food changes, such as those that occur during ripening, storage, curling, rotting, etc. Color is a characteristic of light that may be quantified in terms of wavelength and intensity. It covers all feelings induced by the functioning of the human eye. When light reaches the retina, it stimulates the nerves. Other physical elements may also be examined while analyzing the quality of a meal. In baked goods, volume is vital, while in meat feasts, the texture is essential.

Meat suppleness is an important quality indication. Specific gravity measurements can detect milk that has been diluted. Chemical composition the type and quantity of a food's ingredients determine its nutritional worth and other qualities. In the same manner, as identification and purity standards are created and performed, food safety regulations are based on chemical composition, including chemical pollutants. Proximate analysis refers to the study of the molecular composition of foods. It is used to determine the protein, fat, carbohydrate, ash, and water content of a specific meal (Happiness et al., 2011). Moisture content numerous meals are mostly constituted of water (moisture). In hydrolytic processes, water functions as both a direct reactant and a chemical reaction mediator. By raising the quantity of common saltwater, a direct reactant in hydrolytic processes, water may be extracted or bound from food. Therefore, removing water from food or binding it with common salt or sugar limits different processes and microbial multiplication, increasing the shelf life of various commodities. By reacting physically with proteins, polysaccharides, lipids, and salts, water contributes greatly to food texture.

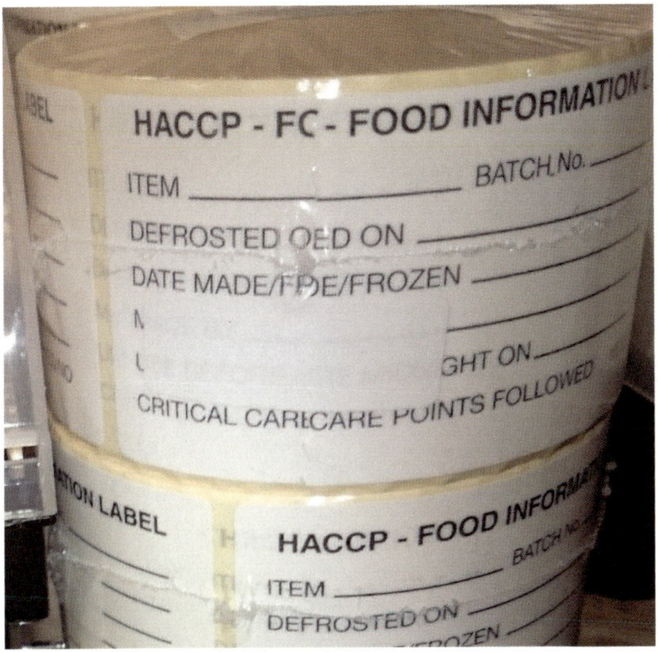

Figure 2.11: Food labeling according to HACCP.

Source https://www.abbotts-catering-supplies.co.uk/HACCP-FOOD-INFORzATION-LABEL-per-500/product/LHACCP/LHACCP/.

The composition of fat the bulk of lipids are synthesized from fatty acids. In these so-called acyl lipids, fatty acids are present as esters, whereas, in a few minor lipid groups, they are present as amides. The acyl residue affects the hydrophobicity and reactivity of acyl lipids considerably. Some lipids are employed to build the membranes that envelope cells and subcellular particles. These lipids are present in the diet, although they represent less than 2% of the total. Nevertheless, even as minor food components, they demand extra care due to their intense reactivity, which can have a major impact on the sensory quality of the meal. Triglycerides constitute the bulk of fats present in animal tissues and plant organs. Such storage tissues can contain 15 to 20% or more lipids, making them an economically viable source for triglyceride separation. This lipid can be consumed as edible oil or fat once it has been purified. The relevance of lipids as fuel molecules (39 kj/g triglycerides) and a source of vital fatty acids and vitamins support their nutritional and physiological worth (Gunaratne et al., 2019). Apart from their functions, lipid characteristics are crucial in food handling and processing. These properties include a pleasant creamy or greasy mouthfeel as well as the capacity to solubilize a wide range of food taste and fragrance components. These attributes are required for food to have the right texture, mouthfeel, and flavor, as well as adequate aroma retention. Furthermore, several goods are deep-fried, which requires frying the meal in a heated fat or oil. Some main food fragrance molecules or precursors decompose, releasing scent chemicals needed as food emulsifiers, while others act as fat-soluble or oil-soluble pigments or food colorants.

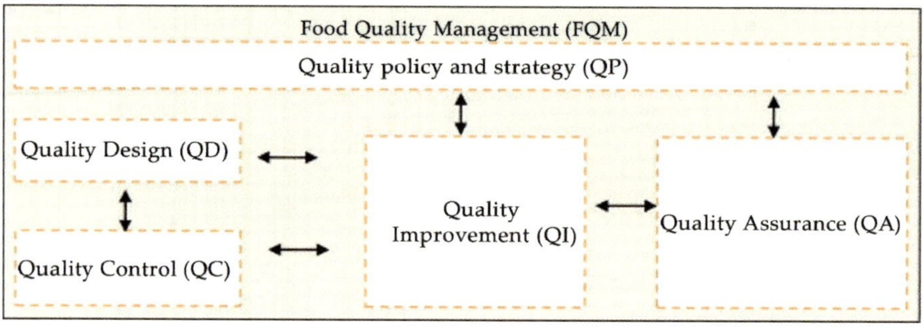

Figure 2.12: Quality management foods.

Source: https://www.sciencedirect.com/science/article/abs/pii/S0956713512006585.

The objective methodology defines the following phases in the approach to quality testing with tools: A physical examination is required. A product's physical attributes are used to judge its quality. A tool-based approach to food quality evaluation can assist in revealing the meal's hidden elements or quality attributes. In general, a sensory test that evaluates quality attributes has a high correlation with a quality-evaluation instrument. The physical test measuring technique evaluates color, volume, texture, viscosity or viscosity and consistency, pliability and tenderness, and specific gravity. Advantages: It has a significant impact on the final product's quality (Fiorentini et al., 2020). This approach allows you to rapidly and conveniently get the results of measurements and observations. Can help with the analysis of a company's productivity or marketing activities.

Deficiency: Some sensory properties impose constraints that cannot be stated in language. Tools and instruments must be calibrated regularly to ensure that basic tools are used appropriately and that their outputs are accurate. Miscommunications between management and panelists are also conceivable. Physical testing devices and apparatus include the spectrophotometer, planimeter, texturometer, penetrometer, and hydrometer. There are lactometers and alcohol meters available. Saccharometer. Applied mathematics: A chemical test measurement process based on a product's chemical composition can be used to determine its quality. Proximate analysis, which measures water and ash content, and qualitative/quantitative analysis, which measures protein, fat, carbohydrate, fatty acid, reducing sugar, and amino acid content, are the two types of chemical testing processes. One advantage is the employment of very objective methodologies. The findings are reliable (high reliability). The chemical nutrients included in a portion of food may be able to evaluate its quality (Feng et al., 2019). One disadvantage is that it is expensive. Complex. Chemical analysis expertise is in high demand. The use of chemical reagents necessitates a high level of precision and safety. Analytical procedures involve a wide range of techniques, such:

Gravimetry Sediments are extracted, weighed, and examined in Gravimetry. Gravimetric analysis is a weighing-based quantitative analytical approach. The separation of the components of interest from the remainder of the sample is the first stage in gravimetric analysis, followed by deposition. In the gravimetric technique, the relative atomic mass, relative molecular mass, and precipitate created by the reaction all contribute to estimating the number of analyzed components. Gravity analysis can be performed via precipitation, evaporation, or electrolysis.

2.5. VOLUMETRY

Volumetry is an analytical technique used to measure the concentration of a chemical by calculating how much solution is required to quantitatively react with another chemical solution, such as a known concentration of another drug. The outcome is determined by titrating in volumetry, a technique that adds known concentrations of a standard solution (in the form of burettes) to a known or titrated solution until they react entirely and achieve the matching volume of solution. Titrations are completed when the standard reaches zero, which equals the solution being titrated. Indicator chemicals are used to determine how much of the standard solution has interacted with the titrated solution. Titrant addition can aid in determining when it should be discontinued (Elsamani et al., 2014). When the reaction is finished, you should observe a distinct difference in the look of the drained solution and the standard solution (e.g., in the presence of a change in color or the formation of a precipitate). When the indicator changes color, the titration is complete and must be stopped. The two types of standard solutions utilized in volumetry are the primary standard and the secondary standard.

The capacity to rapidly determine numbers based on weighing findings is the most significant criterion. The most used unit of measurement is mole equivalents per liter, or M. (mole equivalents per liter). The major standard solutions are NaCl, oxalic acid, and sodium oxalate. The concentration of secondary standard solutions can be determined using freezing, main standard solutions, or gravimetric techniques. As an example, consider NaOH (standardized with oxalic acid primer). The raw material must meet the following requirements: Its chemical composition is undeniably well-known. When it comes to housecleaning, keep it straightforward. Because it is non-hygroscopic, drying is an option. Both its pure and dissolved forms are stable. Before reacting with the sample, a stochiometrically calculated standard solution or substance can be distributed in a suitable solvent. The comparable weight is important to reduce the impact of weighing errors.

2.6. SPECTROPHOTOMETRY

Spectroscopic techniques such as prism monochromators and diffraction grids, as well as detectors such as vacuum phototubes and vacuum photon tubes, can be used to quantify the absorption of monochromatic light by an array of colored solutions. Spectrophotometry might be used as a more complete visual inspection that focuses on energy absorption. Light absorption by a material is measured at various wavelengths and processed per CaM to provide spectra for each component (Elina et al., 2016).

Chromatography is a molecular separation technique for separating solution components by altering the patterns of movement between the mobile and stationary phases (in the form of molecules). The dissolved molecules will pass via a column into the stationary phase after dissolving in the mobile phase. A molecule with a high affinity for the column travels slower than one with a low affinity. Column mobility can be used to distinguish between various types of molecules. Components eluting from the column can be identified using a detector or collected for future research. In two-phase differential chromatography, the stationary phase and mobile phase are dispersed differentially, as previously stated.

The movement of the mobile phase in the reaction chamber causes differential migration of sample components. Bacterial testing A total plate count (TPC), for example, counts the number of bacteria, mold, yeast, and protozoa present. The microbiological test may be used to assess a product's shelf life as well as the cleanliness and safety of food. The quality and longevity of a meal may be assessed using qualitative tests, quantitative testing for pathogenic bacteria to determine its degree of safety, and indicator bacteria test to determine the level of food cleanliness (Ebuehi & Oyewole, 2007). Microbiological tests are performed on a product to objectively assess its quality based on the presence of microorganisms. Despite its use in determining the cleanliness of food sanitation, this test has several limitations, including the risk of infection among test volunteers and the time-consuming procedure of culturing bacteria.

CHAPTER 3

CHALLENGES IN ENSURING CONSUMER ACCEPTABILITY AND MARKET SUCCESS OF PRODUCTS

CONTENTS

3.1. Functional Challenges in Ensuring Consumer Acceptability and Market Success of Products 54

3.2. Other Functional Challenges a Company Must Confront to Ensure Market Success of a Product 58

3.3. Psychological Challenges 59

3.4. Economic Challenges 63

3.5. Sociological Challenges 68

3.6. Environmental Friendliness 73

3.7. Product Quality and Standards Take Center Stage in Consumer Choice 75

The most basic challenge that every new product faces is consumer acceptance. The company must ensure that the new product will be able to meet consumer needs and wants and also satisfy other criteria of consumer acceptability. If the product is acceptable, it has a chance of being preferred over competing brands. But if the product is not even acceptable, success may elude it altogether, no matter how well it is marketed (Yağcı & Göğüş, 2008).

Acceptability is defined as the extent to which a person or group of people find a particular product or service consistent with their needs, wants, and preferences. Acceptance can be influenced by many things. Product success depends on a well-planned strategic marketing program that addresses several key issues: the definition of product positioning, which may require considerable market research; the development of appropriate marketing communications (e.g., advertising) that is consistent with product positioning; and the selection and implementation of distribution channels designed to maximize sales volume.

Figure 3.1: Customers are the main influencers and pillars of a business. Their entitlement to the success of a product is key.

Source: Image by Qminder.

In today's global economy, products must be designed and manufactured to appeal to the tastes of consumers around the world. Being able to create products that are accepted by foreign markets can make or break a business. Companies need to be able to adapt their products and services so that they are accepted by foreign cultures. The market success of a new product depends on many factors, including the product's suitability for its intended use, its technical qualities and other attributes that affect performance, price, appearance, reliability and availability (Yang & Boyle, 2016). Consumer acceptability is an important element in ensuring market success. Market researchers aim to understand how consumers perceive a product so that they can predict its commercial success by assessing whether or not the product will perform well in stores. In order to achieve this objective, they must perform many tasks – some simple and straightforward but others complex and difficult.

There are many challenges in ensuring consumer acceptability and market success of products. The product must be developed for the intended use, which requires appropriate testing and product design. The resulting product must be economical to produce and be durable, safe, and attractive. It must be available at a reasonable price, either through mass production or by providing a unique product that is worth paying a premium for. When introducing a new product onto the market, various obstacles will have to be overcome in order to ensure consumer acceptability and market success. These challenges can be classified into four broad categories: functional, psychological, economic and sociological (see table below). The functional aspects of new product design and development are all those factors related to its functionality since this is what makes it truly different from existing products on the market.

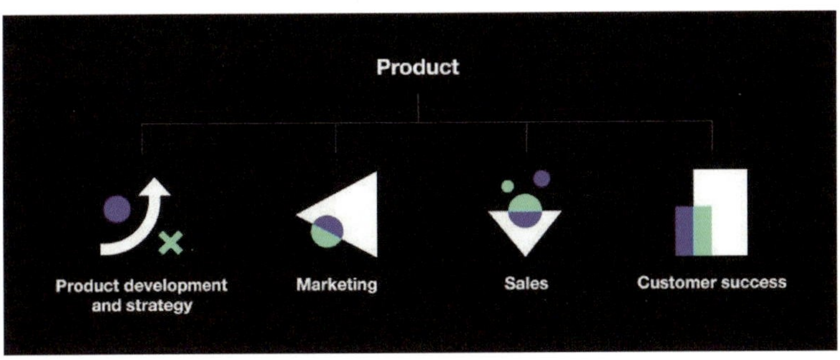

Figure 3.2: In the effect of product marketing, customers highly determine the success of a product.

Source: Image by Business 2 Community.

Psychological challenges relate to factors such as how consumers perceive new products in relation to already accepted products, how they evaluate the need for their existence and how they evaluate their performance potential.

Economic concerns relate to profitability issues resulting from differences in price-quality perception at different levels of demand, as well as uncertainty regarding possible price movements within the markets of newly introduced products (Warnakulasooriya et al., 2020). Finally, sociological factors are all those issues related to society's acceptance of new products with respect to cultural differences across national boundaries (e.g., language barriers) or religious taboos.

3.1. FUNCTIONAL CHALLENGES IN ENSURING CONSUMER ACCEPTABILITY AND MARKET SUCCESS OF PRODUCTS

3.1.1. Speed to Market

Speed to market is the time it takes for a product to go from initial idea to being in the hands of consumers. This is an important factor because it can have a tremendous impact on how well your product does in the marketplace and whether or not you can sell it at all. The bigger issue with speed to market is that you may need to spend a significant amount of money on research and development before you even know if your idea has any merit (Tsai et al., 2021). You could end up wasting resources on something that no one wants.

Figure 3.3: Speed to market is not only a matter on how and when a product is to be delivered but an argument of factors that lead to the delivery of a product.

Source: Image by CBX.

3.1.2. Quality Assurance

Quality assurance is the process of ensuring that a product is of the right quality. This can be achieved through testing, inspection or other means. Quality assurance includes all the activities that ensure this happens, including:

- preparing specifications for design and construction;
- testing during manufacturing to make sure parts meet required standards.

Figure 3.4: The 5 major steps involved in quality assurance.

Source: Image by Sofeast.

In addition to these two main areas, quality assurance also includes inspecting finished goods before they're shipped out; checking incoming raw materials against specifications; conducting incoming inspection on items received from suppliers; and inspecting packaging materials before they are used in any packaging process (Tuorila & Monteleone, 2009).

3.1.3. Innovation

Innovation is a major challenge that faces the consumer acceptability and market success of a product, service or idea. Innovation is an ongoing process which requires constant attention to keep up with customer needs and demands. As such, it requires continued innovation in order for a company or brand to remain relevant and competitive in the marketplace. Innovation enables companies to differentiate themselves from competitors through new products, services or ideas. It creates a unique value proposition that makes customers want to buy something they don't yet have (or have never experienced). Innovation also helps create customer loyalty by making them feel special; like they're part of an exclusive club—and will lead them back again when they're ready for another purchase!

3.1.4. Regulatory Compliance

Regulatory compliance is the rules that govern the use of your product. Regulations can be federal, state or local and often vary depending on the industry. For example, a regulation that requires companies to file an annual report with the Securities and Exchange Commission (SEC) might not apply in your industry. Regulatory compliance can be specific to your company as well; for example, you may need to comply with U.S. Food and Drug Administration (FDA) regulations but not those from other agencies such as OSHA or EPA.

Figure 3.5: In breach of the regulatory compliance, a company faces the threat of losing its customers.

Source: Image by Thomasnet.

In addition to these general regulatory guidelines, there are regulations that apply specifically to consumers as well. The FDA regulates foods and drugs sold in the United States through its Center for Food Safety & Applied Nutrition (CFSAN), which oversees more than 30 different offices including food safety inspection services at all domestic ports; seafood identification programs; seafood quality control programs; food additives approval procedures; new animal drug applications; dietary supplements review program management activities; postmarket surveillance activities related to human drugs (including pharmaceuticals); medical device registration fee collection activities; among others (Short et al., 2021).

3.1.5. Corporate Sustainability

Corporate sustainability concerns the company's ability to sustain its business practices in the long term. In order to be sustainable, companies must consider the environmental, social and economic impact of their operations (including their supply chain). This type of thinking requires a different approach than what has traditionally been used by businesses.

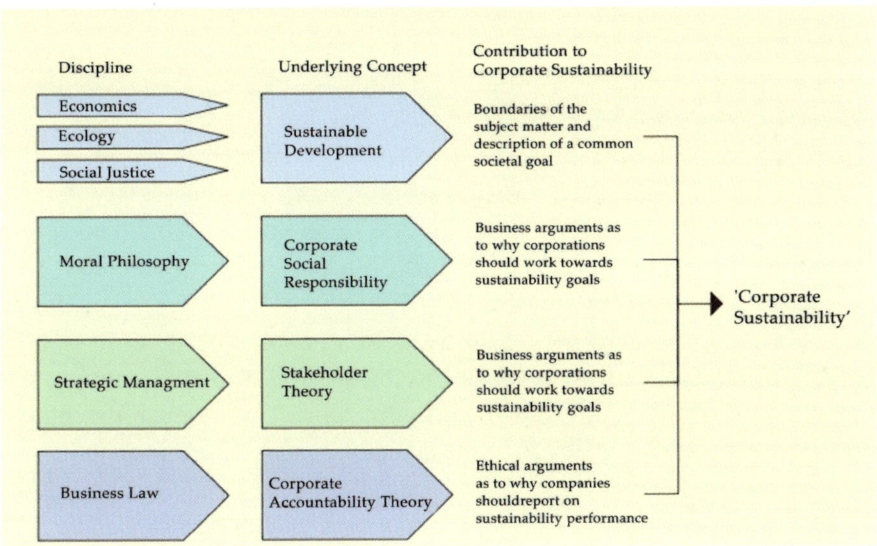

Figure 3.6: The various disciplines of corporate sustainability. Corporate sustainability is a challenge in modern day business setting.

Source: Image by Ivey Business Journal.

In order to succeed as a sustainable brand, brands need to understand consumer needs and how they can use their products as tools for addressing these needs. For instance, if sustainability is a key consumer need then this should demonstrate itself through consumer behavior such as purchasing environmentally friendly products or avoiding unnecessary packaging when buying groceries. Furthermore, if consumers believe that companies are not doing enough with regards to sustainability then they could choose another brand which better supports this belief – even if both brands offer similar products!

Corporate sustainability is a challenge in ensuring consumer acceptability and market success of products. In this sense, it is the corporation's responsibility to anticipate changes in trends, consumer preferences and expectations. The purpose of this thesis is to analyze how a corporate social responsibly program can influence brand reputation, consumer behavior and corporate image through the development of its CSR policy. As a result of this study we find that a strategy that is based on corporate social responsibility has an impact on brands' reputation and consumers' behavior as well as on companies' image (Sutrisna et al., 2017).

Since the early 1990s, companies have increased their investments in corporate sustainability. This study explores how corporate sustainability is related to consumer acceptability and market success of products in different product categories. It uses two waves of data from an international retail company as well as two waves of data from a large survey study for a month and a half period between fall 2007 and spring 2008. The findings suggest that corporate sustainability has both positive and negative effects on consumer acceptability, particularly for low involvement products with high visibility. For example, initiatives to reduce carbon emissions are positively related to both over all acceptability and relevance ratings, while initiatives to improve safety are negatively related to overall acceptability but positively related to relevance ratings.

3.2. OTHER FUNCTIONAL CHALLENGES A COMPANY MUST CONFRONT TO ENSURE MARKET SUCCESS OF A PRODUCT

One of the first challenges facing companies considering new products is to ensure that they can deliver these products in a timely manner. In many cases, this means meeting an aggressive deadline for launching a product or service that will be offered at a lower price than competitors' offerings. To meet this challenge, companies must prioritize speed over quality assurance and innovation by not spending time testing their products in the market before launch. In addition, they may have to cut corners on regulatory compliance or brand standards in order to accelerate development timeframes.

The increasing importance of corporate social responsibility, and congruity with consumers' changing expectations are just two of the issues that need to be addressed by companies wishing to capture market share by selling sustainable products.

3.2.1. Environmental Responsibility

Environmental responsibility is becoming more and more important, particularly to consumers in developed countries, and is an area in which many companies are working hard to improve. Environmental concerns are a significant challenge for product manufacturers because the environment affects everything they do, from raw materials acquisition through manufacture, distribution and disposal of products.

It is important for businesses to consider environmental responsibility as a challenge in ensuring consumer acceptability and market success of products. In a world of increasing resource scarcity and environmental awareness, the importance of environmental responsibility has been recognized. Acknowledging this, companies do conduct extensive research into raw materials sourcing, sustainable development and recycling (Schiling & Pham, 2012). Successful implementation of these initiatives enables them to provide environmentally responsible building materials that enhance quality of life and allow our customers to leave their footprint on the world in a positive way.

3.3. PSYCHOLOGICAL CHALLENGES

3.3.1. Hesitancy to Accept New Products

The first challenge is the hesitance to accept new products. Consumers are reluctant to accept a new product because it is not well-known and cannot be trusted. People do not want to waste money on an unknown brand or try something they have never tried before, so they stick with what they know. They are often unwilling to change even if the new product offers better quality or value than existing ones.

3.3.2. Difficulty in Getting Products Tested

A product can be developed and put into the market for sale only after it is tested by a neutral person. The testing should be done in a scientific way to ensure that the product meets the expectations of consumers and also complies with relevant standards, laws, and regulations.

Figure 3.7: The reluctance in getting products tested turns out ravaging for institutions and business out of the unexpected delivery other than the consumer's needs.

Source: Image by poll the people.

The following questions need to be addressed before conducting a market test of any product:

- Who will perform this test? A group of people or an individual?
- Will the testing take place on one occasion only or will it involve multiple trials over time?

For example, if you want to test whether your new line of hot sauce contains enough spice to make it popular with customers, you could conduct one trial during summer or another time when they usually eat spicy foods (e.g., around Thanksgiving). You could even conduct several trials under different weather conditions such as heat or coldness depending upon where your customers live and what type of activities they do outdoors during those times (such as running marathons) (Shinde et al., 2016). This would help ensure that your product works well regardless if someone uses it inside versus outside their home—or somewhere else entirely!

3.3.3. Lack of Vocabulary in Talking About a New Product

One of the most common challenge is that it is difficult to describe a new product, even when you have the best vocabulary and experience at your disposal. The words we use to describe products tend to be static and are often too technical or complex for people who don't know what they mean. We try to make things easier by using metaphors, but this can backfire if the metaphor isn't familiar enough for your audience:

Figure 3.8: Poor vocabulary tend to veer away potential customers.

Source: Image by MEA ENGLISH.

For example, "This new product will revolutionize how you clean your kitchen surfaces" sounds better than "This new cleaning solution removes 99% of germs from kitchen surfaces in just 30 seconds!" But if someone has not heard about cleaning solutions before or doesn't understand what "germs" are then they will probably feel confused and uncomfortable rather than excited by the prospect of their kitchen being cleaner than ever before (Sun et al., 2021).

3.3.4. Subjectivity in Perceiving Quality and Utility of a Product

There are many factors that influence the perception of quality and utility. These factors include, but are not limited to:

- The price of the product.
- The expectations of the consumer about this type of product, as well as his/her previous experience with similar products from different companies or brands.
- The environment in which the product is used (e.g., how easy it is to use and maintain).

In addition to these external influences on consumers' judgments about a specific product, there are also internal factors that affect their perceptions of quality and utility.

3.3.5. People Respond Differently To Different Products

A common mistake marketers make is assuming that everyone has the same needs and wants, and therefore they would all be equally receptive to a given product. While it's true that all of us have basic needs (e.g., food, water), these needs can vary greatly depending on our age, lifestyle, culture and so on. In addition to these factors contributing to how we perceive products as consumers look for brands which will meet their specific needs; every person has different tastes and opinions when it comes to what they like or dislike about a product.

For example: you may prefer seafood while your friend likes meat dishes only; your sister prefers to read romance novels while your brother likes sci-fi stories more than anything else; you might prefer luxury cars while other people might think twice before buying them due to their high price tags etc. (Serventi et al., 2022).

Marketers need not just understand these differences but also leverage them during marketing campaigns when designing new products so as not only appeal directly towards potential buyers but also differentiate themselves from competitors who produce similar products or offers at lower cost points (e g competitors offer similar services at lower prices).

3.4. ECONOMIC CHALLENGES

3.4.1. Research and Development

In order to develop successful products, it is essential that you understand the needs and demands of your customers. You should also be aware of what the competition is doing in terms of product development. In addition to this, it's important to keep track of legal requirements and regulations regarding the industry you are operating in. And finally, you must understand marketing factors such as promotion pricing and advertising strategies so as to achieve maximum sales for your product at minimal cost

Figure 3.9: To understand what the consumer needs is essential in the market success of a product. This explains the need for research and development.

Source: Image by Inc. Magazine.

3.4.2. Product Design

- The product design should be designed for the lowest common denominator.
- The product design should be designed for the average person in the market.
- The product design should be designed for the upper end of consumers in your target market.
- The product design should be designed for a niche market segment of customers who will pay more than usual because they

value certain attributes highly, such as low emissions or recycled content (Silva et al., 2020).

- The product design needs to appeal to customers who have an above average purchasing power, but at a reasonable price point that does not make it inconvenient to use their products on a day-to-day basis (e.g., $100 per month).

3.4.3. Legal and Regulatory Requirements

Legal and regulatory requirements have the potential to significantly affect consumer acceptability and market success of products. As a result, it is important for companies to understand legal and regulatory requirements at both national and international levels. A key challenge facing companies is that legal and regulatory policies can be complex, rapidly changing and difficult for non-experts to understand. They are often seen as major barriers to a successful market introduction. However, if considered early in the product development process, legal and regulatory requirements can be planned for, approached in a structured manner and mitigated (Rousta et al., 2021). This is essential to ensure that products meet every internal and external requirement, such as usability, safety or environmental impact.

A number of legal and regulatory requirements can affect the acceptability and success of a product. These include:

- Product safety laws, including those that require products to be fit for purpose, safe and durable;
- Product liability laws which impose obligations on manufacturers to take reasonable steps to avoid harm caused by their products;
- Competition law rules that prohibit anti-competitive behavior such as price fixing or market sharing agreements between competitors; and
- Environmental protection legislation which imposes obligations on producers in relation to waste disposal and pollution control.

3.4.4. Marketing Aspects (Consumer Acceptability)

Marketing aspects are deemed to be critical to the success of any product. This is especially so when it comes to Public Health products and services. There are standards in place that have been developed by national and international bodies, but they are often time difficult to implement. Taking a decision regarding the acceptability of a product should be based on research findings, not on subjective assessments (Pu et al., 2020). Acceptability will then be translated into the demand for a product, which will affect its market success.

Figure 3.10: The four key As of marketing that influence consumer acceptability.

Source: Image by marketing 91.

Consumer acceptability is one of the important factors in the success of a product. Successful products are those that are accepted by consumers and used regularly. Consumer acceptance can be measured using various methods, including surveys or focus groups. It is also useful in predicting market potential for new products, as well as assessing changes to existing products or services. Consumer acceptability can help companies develop:

- Products with higher levels of suitability for certain populations (i.e., older adults).
- New ways to promote their products (e-commerce sites, social media).

3.4.5. Distribution Channels for the Final Product

In this case, you will need to determine the best distribution channel for your product. There are many different types of distribution channels that a company can use to sell its product, such as direct sales or third-party resellers. The method of distribution depends on the type of product and the size of operation. If you choose direct sales, then this means that you will be selling your goods directly to consumers or businesses and there would be no intermediaries involved in the process (Peinado et al., 2014). A perfect example is Amazon's marketplace where products are sold by sellers who have their own inventory but also can use Amazon's fulfillment services if they don't have any warehouse space themselves.

3.4.6. A Number of Critical Elements Must Be Considered

A number of critical elements must be considered when delivering a product to the market. These include:

- Design – The design of a product will affect its functionality, reliability, durability and cost. It is important to consider how people are likely to use the product (e.g., whether it is used in an office environment or outdoors).
- Legal and regulatory requirements – The marketing aspects of new products often focus on consumer acceptability; however, legal and regulatory requirements may also play an important role in ensuring market success for new products. It is essential that you understand these issues before proceeding with your marketing strategy for your innovative product/service (Perera et al., 2014). For example, there may be restrictions on labeling or advertising claims relating to health benefits from consuming certain foodstuffs; or there may be obligations under intellectual property laws that prevent you from making misleading statements about third party products within your own advertisements (for example where there are trademarks listed).

3.4.7. Sensorial and Non-Sensorial Attributes

There are two types of product attributes:

- Sensorial attributes are those that can be experienced through the senses and include (but are not limited to) color, smell, taste and texture.
- Non-sensorial attributes cannot be experienced through the senses and include (but are not limited to) durability, ease of use and value for money.

Often associated with the physical properties of a product or service, sensorial attributes may have an impact on consumer perception when deciding whether or not they will purchase it.

3.4.8. Economic Hardtimes

The economic hardtimes have prompted consumers to look for value-for-money products. Hence, they are more likely to purchase a product that offers good value for money than one that is expensive. This implies that manufacturing companies should be able to meet the expectations of their target consumers by providing products that offer adequate quality at affordable prices.

Social and political factors can also influence consumer acceptability of a product since some consumers may decide not to buy a particular brand because it is manufactured in a country whose government is not favorable to them or their beliefs, regardless of its intrinsic quality or price competitiveness. Sensorial and non-sensorial attributes: Many people judge how well something works by looking at its appearance rather than its performance on an objective scale (sensorial attribute). Some other aspects like odor, taste, texture also influence people's perceptions about whether food items such as breads/biscuits etc., will taste good – even if they do not affect directly any functional properties of these food items (non-sensorial attributes) (Predieri et al., 2021). Therefore, all these sensory characteristics must be considered while designing new products so that they have high consumer acceptability rates which translate into profitability for manufacturers once they enter the market place successfully through distribution channels established beforehand

3.4.9. Political Boycotting

The boycott is a powerful tool in the hands of consumers who want to express their dissatisfaction with products they feel are not socially responsible, such as those that use child labor or fail to provide a good working environment for employees. It can be applied to any product, service or company anywhere in the world because it is an economic sanction against an entire country or region—and not against specific individuals or companies.

Figure 3.11: Political boycotts affect the market success of a product.

Source: Image by BBC.

Boycotting works by reducing demand for products from companies that engage in unethical practices and increasing demand for ethical ones. When people stop buying from unethical companies, those companies lose profits and eventually change their ways so they can make more money again.

3.4.10. Positive Disconfirmation

There are various ways of doing this, but one of the most common is known as positive disconfirmation. Positive disconfirmation is when you test a product with consumers to see if they will accept it and whether it changes their behavior in any way. For example, if you're developing a new toothpaste or shampoo and want to know if people will buy it, you might perform positive disconfirmation by getting people to try your product first and then asking them questions about it afterwards (O'Mahony, 2018). The results would tell you how well the product fit what people wanted (or didn't), and whether they'd be willing to purchase more in the future.

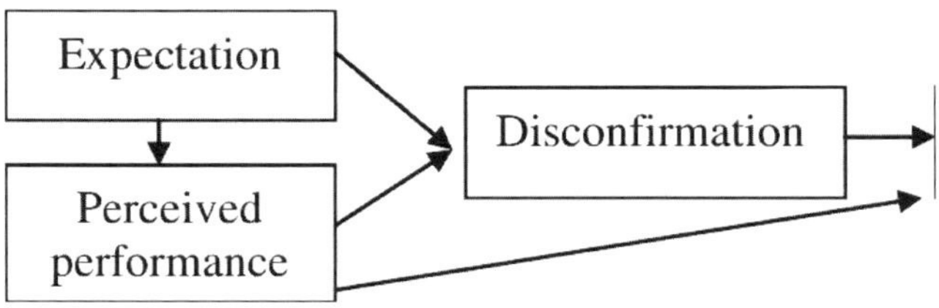

Figure 3.12: The theory behind positive disconfirmation.

Source: Image by Researchgate.

3.5. SOCIOLOGICAL CHALLENGES

One of the challenges in ensuring consumer acceptability and market success of products is making them attractive to customers. A successful product must meet the needs of consumers, who are always looking for better ways to live their lives or run their businesses. The developer has an obligation to create a product that's unique and innovative, not copy what others have done; this will increase its chances of success. Another challenge is adapting it so that it keeps pace with current technologies; if a developer doesn't do this, then his product may not be able to compete successfully against other similar ones on the market. Finally, companies need strong marketing strategies in order achieve maximum profits by targeting customers who have both the need and ability to purchase these products at specific prices over time periods (i.e., through promotions).

3.5.1. Innovation and Development across All Divides in the World

The pace at which the world is changing means that there are new opportunities for developers in every industry. The key challenge is to identify these opportunities and develop products that meet consumer needs. The world has become more connected than ever before, with more people living longer and healthier lives while being more aware of their environment and impact on it (Nishinari & Fang, 2018).

3.5.2. Consumer Wants and Needs Evolve As Well

Consumer wants and needs evolve as well. A product that is well accepted today may become obsolete in a couple of years due to changes in the lifestyle, culture and economy of the society. The need for new products to meet the changing needs of consumers is continuously increasing for two reasons: firstly, it helps in meeting the competition from other companies by providing better services; secondly, it increases sales revenue for businesses by making them more competitive than their competitors.

3.5.3. The Technology behind the Product Must Be Unique and Innovative

Another common challenge is the technology behind the product. The technology should be unique and innovative enough to differentiate it from other products in the market. In order to ensure consumer acceptability, a successful product must be based on something that is both new and different from all other existing products in its category. New technology has been defined as "any invention that does not exist today or which represents an advance over existing inventions" (Miller et al., 2005).

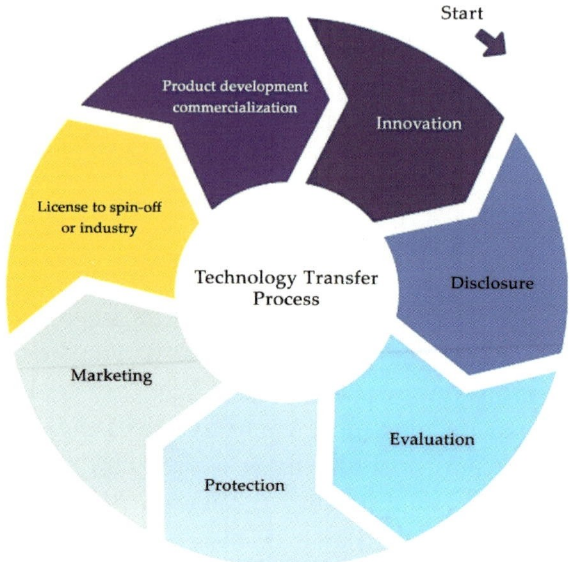

Figure 3.13: Technology is all about commercialization.

Source: https://www.utoledo.edu/research/TechTransfer/TTandCommProcess.html.

Innovation refers to "the process of introducing new ideas, services or goods into society." Innovation begins with an idea or need that can be turned into a better product by solving problems or meeting unmet needs of consumers. It requires creativity in thinking outside of the box while applying modern science so as to create something useful for people around us every day.

3.5.4. If It Simply Copies What Others Have Done, There's Little Chance It Will Succeed

It is important to note that the successful product must be different from, and better than, what is already on the market. In fact, it's best if your product offers something unique and innovative, whether in features or design. For example, in the case of audio equipment for automobiles (e.g., stereos), a new car stereo may have some advanced technical features such as an LCD display screen and Bluetooth capability (Nwakalor & Obi, 2014).

At this point you might be asking yourself: If everyone else has a stereo with these same features already, why would anyone buy my product? There are several reasons why someone would choose one model over another—for instance:

- Brand recognition. Perhaps they like Sony products better than other brands; or maybe they would prefer choosing an Alpine model because they're more familiar with that name brand than others available on the market today; or perhaps there is no brand loyalty involved at all—people simply want whatever works best for their needs at any given time (and not necessarily just based upon price).
- A product may also require modification to bring it up to speed with current technologies; if it doesn't add value to a person's life or business, it won't become popular or generate sales.

One of the challenges faced by companies is keeping up with current technologies. Developing a product that is innovative and can be marketed successfully requires constant monitoring of trends in technology and consumer preferences. As technology changes at a fast pace, it is inevitable that older products will eventually become obsolete. A company may also require modification to bring it up to speed with current technologies; if it doesn't add value to a person's life or business, it won't become popular or generate sales.

3.5.5. Marketing Strategy

A strong marketing strategy is the key to success. A strong marketing strategy is a plan for promoting a product or service to a specific target audience. The marketing strategy should be based on research and data about the target audience, as well as the product itself and its competitors.

For example, if you want to sell a new flavor of potato chip that has more cumin than any other chip on the market, your marketing strategy will likely focus on your target market of people who love cumin and are willing to pay more for it. If you want to sell gardening supplies for people who have no experience with gardening but love cats and dogs, then maybe your best bet would be advertising on Instagram because these customers are likely young women who spend most of their time online looking at cute animal photos (Muhimbula et al., 2011).

3.5.6. Target Consumers

- It must target the right group of consumers who have the need and ability to buy the product.
- The marketing plan must be able to reach and influence those people with reasonable effort and cost.
- The message you are sending should be consistent with your target market's expectations, values, beliefs, attitudes, motivations, and needs.

3.5.7. Impact of Advertisement

For example, a successful soft drink company advertised its beverages by pointing out that they can help people regain their energy after a hard day at work. This is an excellent way to address consumers' needs and wants because:

- It appeals directly to the consumer's personal experience; and
- It focuses on what the consumer wants from the product.

3.5.8. Response to Consumer's Preference

The company also releases new flavors periodically in response to consumer preferences, which keeps customers coming back for more. The company is always looking for ways to improve the product and make it better. For example, they recently released a new flavor and are now in the process of evaluating its success.

3.4.9. Desire for Value

Consumers want value for their money and reward companies that provide it. This is true whether they're buying a new pair of pants, a car, or the most advanced product on the market. Just because you've developed something that's better than what's currently available doesn't mean people will buy it if it costs too much in comparison to other options (Munasinghe et al., 2013). To sell more products and make more money in the long run, you have to understand what consumers want from your product and price accordingly. The challenges in technical aspects include:

1. **The aspect of safety:** Safety is an important aspect of consumer acceptability and market success. The safety of a product is a key factor in its sustainability, profitability and market success.

2. **The quality of affordability:** As a key factor in the acceptability of a product, affordability is a function of price and income. A low-quality product that is more affordable than its high-quality counterpart will be preferred by consumers if their incomes are low. On the other hand, if the quality of two products is identical but one costs more than the other, consumers with higher incomes will choose the lower-priced option.
3. **The quality of availability of supply:** The quality of availability of supply is a key factor in consumer acceptability and market success of products. In many countries, especially in the developing world, there are chronic shortages and/or poor quality of basic goods such as food, clothing and housing. This may be due to a variety of reasons: lack of adequate financial resources; natural disasters; political instability; poor management systems; poor transport infrastructure, etc. (Mouta et al., 2016). However, it is generally accepted that these factors derive from an inadequate supply response by producers or traders to demand pressures generated by population growth or increases in real incomes.

In this context, there are several reasons why firms might fail to produce more when they can make higher profits from doing so:

- They may have limited ability to ramp up production rapidly because they have not invested sufficiently in capacity expansion before now (in other words "capacity constraints"). The classic case here would be firms who think that experience curve effects mean that larger volumes will lead automatically or at least predictably to lower costs per unit produced

3.6. ENVIRONMENTAL FRIENDLINESS

Environmental friendliness is a quality that can be quantified, marketed, promoted and sold. This means that environmental friendliness is not only an important factor in ensuring consumer acceptability and market success of products but also that it has the potential to become a new source of revenue for companies.

However, companies need to be careful when pursuing this new source of revenue because if they do not understand what makes people buy products with environmentally friendly qualities then their attempts at marketing and promoting these types of products could backfire and hurt their business instead (Lomarat et al., 2019).

Figure 3.14: A healthy environment attracts more consumers and makes them accept the product at their convenience.

Source: Image by WWF-SCP.

1. Safety, Affordability, Availability of Supply, and Environmental Friendliness Are Important Qualities to Consider

- Safety. The product should be safe for consumers to use, or at least no more dangerous than other products that consumers are already using.
- Affordability is also important because it determines whether the product reaches a mass market and can be sold at a low price without sacrificing quality (if possible).
- Availability of supply is another factor to consider because if companies do not have enough supplies to meet demand, they will not be able to sell many items and make profits from these sales as well as other aspects of their business such as advertising costs associated with promoting their products through television commercials, billboards etcetera; therefore this would mean that profits could potentially go down significantly if there is not enough supply available so there must be some sort of solution found before any serious damage occurs within an organization's operations (Lee, 2013).

2. In a developing country like Nigeria, the challenges of ensuring the acceptability and market success of products are enormous due to lack of adequate industrial infrastructure and trained manpower. This includes:

- Lack of adequate industrial infrastructure (e.g., factories).
- Lack of trained manpower.
- Inadequate power supply.
- Inadequate water supply.
- Poor transportation system (e.g., roads).
- Poor communication facilities (e.g., telephones).

In addition, there is also a need for effective marketing and research and development activities in order to ensure that a product's acceptability is improved upon before it reaches the market place.

3.7. PRODUCT QUALITY AND STANDARDS TAKE CENTER STAGE IN CONSUMER CHOICE

In the market place, consumers are faced with a plethora of choices and options. The reason is that the product diversity available today has made choices very complex. While most businesses see quality as the key factor in consumer choice, they do not realize that price is also an important determinant of consumer choice. Product availability and safety are also important factors in consumer choice (Ayzeli et al., 2020).

Besides these four main factors, environmental friendliness also plays an important role in influencing customer decision making process when buying goods or services – this has been happening since 1970s when first environmental movements started emerging on world stage.

3.7.1. It Must Be Technically Feasible

The product must be technically feasible to be acceptable to consumers and market success. If the product is not technically feasible, then it's not acceptable to consumers or market success. In today's highly competitive world, every business needs to have a product that resonates with its customers. The success of your company depends on how well you understand consumer needs, identify gaps in the market and develop innovative solutions. In order to design and market products that meet consumer needs and expectations, it is important to conduct consumer research continuously (Abdulmajid et al., 2014). The goal of such research is to identify opportunities for growth while avoiding failures as early as possible by discovering unmet needs or unsatisfied wants in the target audience before launching a new product or service into the market.

3.7.2. Focus on Pain points

- Focus on pain points. The key to a successful product is to understand the consumer's pain point and focus on improving it.
- Ask consumers what their pain points are, so that you can improve upon them. This may seem like common sense, but it's actually not always common practice! A lot of companies assume that they know best, and don't bother asking the consumer what they want or need in a product. Don't make this mistake—it will only lead to frustration for both yourself and your customers!
- Pain Points can vary from person to person. For example: one person might think that price is more important than style when shopping for shoes; whereas another person might think price isn't important at all and instead wants something simple yet stylish (like Converse). In other words: every individual has different needs when it comes down choosing products—and therefore won't necessarily be satisfied by whatever solution someone else might offer them without first getting some feedback beforehand (Abamecha, 2020).

3.7.3. Managing Brand and Market Dynamics with the Changing Times

The market dynamics are constantly changing and companies need to adapt their strategies accordingly. A company that can identify and adapt to the changing market dynamics, anticipate changes in it and respond appropriately will have a competitive advantage over its competitors.

3.7.4. Problems in Customer Satisfaction and Market Success

Problem 1: Poor product quality. It is crucial to ensure that the quality of your product meets customer expectations, as this is often the difference between success and failure. If your customers are not satisfied with the quality of your products, they will be more likely to switch brands in favor of one that offers better value for money.

Problem 2: Poor service delivery. If customers do not receive satisfactory service from you or your representatives when they purchase or use a product, they may grow dissatisfied with their experience and take their business elsewhere in future (Lee et al., 2021). When it comes to providing excellent customer service over multiple interactions throughout the life cycle of a product (i.e., from purchase through disposal), it's important for businesses to adopt a flexible mindset that allows them to continually innovate with new ideas on how best to meet people's needs at every stage.

3.7.5. Choices Made Around the Product

To ensure that a product is successful in the market, there are several factors that need to be considered before launching it. If these factors are not properly addressed in time, you might end up with an unsuccessful launch which could lead to losses for your company.

It is a very difficult task for companies to predict the acceptance of their products in the market. In many cases, companies have had to withdraw products from the market after they have been rejected by consumers. This can be quite costly for the company since it has already spent time and money on product development and manufacturing. Currently, these costs are rising due to increased competition in the market. For better or worse, customers will accept a product or service into their lives and make it part of their habits and behaviors based on various factors.

3.7.6. The Impact of Market Segmentation

Market segmentation is the process of dividing a market into different segments. Marketers use segmentation to identify what they believe are the most profitable groups of consumers within their target audience, and then focus on reaching those groups with marketing messages tailored to their needs and wants. Segmenting markets allows companies to develop strategies and tactics that address specific customer needs in order to maximize sales, while also allowing them to adjust these strategies over time as they learn more about their target audience through market research (Lotfy et al., 2021).

By dividing markets into smaller groups, marketers can tailor their product offerings specifically for each niche market segment (e.g., men vs women) or create products specifically designed around customers' specific preferences (e.g., organic vs non-organic). This increases both consumer satisfaction with a product—because it fulfills more needs than some other available options—as well as demand for that particular item because fewer people will want to purchase something similar from another company if it doesn't meet all of their needs in exactly the same way as your brand does (that could lead you toward becoming "buyer's remorse").

3.7.7. Imperfect Social Networking

Social networking has become a prominent and popular means of building relationships. It has been used as a tool for business-to-consumer (B2C) as well as business-to-business (B2B) communications. Social networking helps to build trust, brand loyalty, brand awareness and brand equity, which are key elements of successful brands.

3.7.8. Poor Interaction between Consumers and the Product Itself

The customer is an integral part of the marketing and sales process. The customer's needs, wants and desires are the basis for product development. Without customers, there would be no products created or sold. Products must meet the needs and wants of consumers in order to succeed in a market place. Because of this interaction between products and customers, you can use it as a starting point when developing strategies for your business or organization.

3.7.9. Uncouth Brand Loyalty

Brand loyalty is a positive attitude towards a brand, which is measured by how much consumers prefer one brand over other brands. It can also be measured by how much consumers are willing to pay for the same brand. Brand loyalty is not just limited to products; it can also refer to services and companies. Brand loyalty also affects consumer buying behavior. The product, price, promotion, place, and people all play a role in influencing a consumer's decision to buy or not buy a product (López-López et al., 2019). You should never underestimate the importance of consumer acceptability and market success. Without these two things, even the best products will not see profit or growth. To keep a company afloat, it's crucial for marketing professionals to have an understanding of their target consumers and how they can be effectively engaged with the brand. We hope that you learned something new today about how companies can do just that!

CHAPTER 4

THE HUMAN SENSES

CONTENTS

4.1. Introduction 80

4.2. Human Sensory Organs 83

4.3. Human Sensory Receptors 85

4.4. Sense Organs Threshold 87

4.5. Human Sensory Neuroscience 89

4.6. The Sense of Sight 90

4.7. The Sense of Hearing 93

4.8. The Sense of Touch 95

4.9. The Sense of Taste 97

4.10. The Sense of Smell 99

4.11. Other Human Senses 101

4.12. Challenges Facing Human Senses 103

4.13. Solutions to the Challenges Facing Human Senses 104

4.1. INTRODUCTION

A sense could be a natural system used by a life form for sensation, the method of gathering information around the world and reacting to jolts. For case, within the human body, the brain gets signals from the senses, which persistently get data from the environment, deciphers these signals, and cause the body to reply, either chemically or physically. In spite of the fact that customarily around five human faculties were known; precisely, locate, scent, touch, taste, and hearing, it is presently recognized that there are numerous more (Kweku Amagloh et al., 2013). Faculties utilized by non-human life forms are indeed more prominent in assortment and number. Amid sensation, sense organs collect different boosts; such as a sound or scent for transduction, meaning change into a shape that can be caught on by the brain. Sensation and discernment are essential to about each perspective of an organism's cognition, behavior, and thought.

In life forms, a sensory organ comprises a bunch of interrelated tangible cells that react to a particular sort of physical boost. Through cranial and spinal nerves of the Central and Fringe apprehensive frameworks that transfer tangible data to and from the brain and body, the distinctive sorts of tangible receptor cells such as mechanoreceptors, photoreceptors, chemoreceptors, thermoreceptors in tactile organs transduce tactile data from these organs towards the central anxious framework, at last arriving at the tangible cortices within the brain, where tactile signals are prepared and translated.

The human senses are the five basic bodily functions that allow us to interact with and interpret our environment. The five senses are sight, hearing, touch, smell, and taste. These senses work together to help humans make sense of their surroundings. We use our senses to find food and water, communicate with others, avoid hazards, and recognize danger. Our perception of the world around us is based on information gathered through our senses (Koli et al., 2022). Each one of us has a unique set of sensory capabilities that allows us to perceive different aspects of the environment. Our experiences can also change how we perceive things; a person who grows up in an urban area may have a very different understanding of what constitutes "green" than someone who grew up in a rural area.

Figure 4.1: An image showing a brief summary of the five common human senses.

Source: https://www.istockphoto.com/es/vector/conjunto-de-cinco-sentidos-humanos-gm1249129926-363975657.

The discovery of the human senses is a fascinating and important event in human history. The first attempt at understanding how humans sense their world was made by Aristotle, who believed that humans could only see in black and white. He argued that our eyes were like cameras, capturing light and turning it into an image, but he did not yet understand that color was actually a perception rather than an object. The next major advancement in this field was made by Alhazen, who developed an early version of the pinhole camera. His work led to a better understanding of optics and helped to prove Aristotle's theory wrong: we do not see in black and white, but rather perceive colors based on the way that light interacts with objects around us. It was the first time that humans had a thorough understanding of how they perceived the world, and it helped them to develop a better understanding of themselves and their place within it. The first sense to be discovered was a sight (Kukovics & Németh, 2014). This happened around 30,000 BCE when humans started wearing eye coverings made from animal skins, which prevented them from having to rely on their other senses to navigate through their environment. This allowed for an increased level of exploration, as well as greater awareness of objects around them. The second sense to be discovered was hearing, which happened around 20–10,000 BCE when humans began using bone flutes to communicate with each other over long distances. The bone flutes could also be used as musical instruments, allowing them to express themselves creatively while remaining connected with others who were nearby.

Figure 4.2: An image showing the discovery of human senses.

Source: https://www.istockphoto.com/es/vector/personalidad-humana-futuro-l%C3%ADnea-icons-set-gm927451650–254427085.

Sensory systems, or senses, are frequently isolated into outside, exteroception, and inside, interoception tangible frameworks. Human outside faculties are based on the tangible organs of the eyes, ears, skin, nose, and mouth. Inner sensation identifies boosts from inside organs and tissues. Inner faculties had by people incorporate the vestibular framework, sense of adjusting detected by the internal ear, as well as others such as spatial introduction, proprioception, body position, and nociception; torment. Encourage inner faculties to lead to signals, such as starvation, thirst, suffocation, and sickness, or distinctive automatic behaviors, such as heaving.

Sight is one of our most important senses. We use it to see where we are going and what is around us. Our eyes are made up of many different parts that allow them to function properly. The cornea focuses light onto the retina at the back of our eye, which contains light-sensitive cells called rods and cones. Rods detect black-and-white images while cones detect color images; red, green, and blue. When light hits these cells they send electrical signals to our brain that help us see color and detail in an image. The smell is another important sense because it helps us identify things around us by their scent or odor (Kohyama, 2020). Our nose has special cells called olfactory receptors that react when they come in contact with specific chemicals in different smells like flowers or food cooking on a stovetop – triggering memories from past experiences with those scents. Touch is another way we perceive our surroundings through touch receptors located throughout our body but concentrated in areas like fingertips where they can sense things like temperature. Hearing is another important sense. It allows us to listen for sounds that may be dangerous or helpful. Hearing can also help you understand what someone else is saying by listening to their voice or words spoken in another language.

4.2. HUMAN SENSORY ORGANS

The human senses are the five ways our bodies receive information. The human sensory organs are the eyes, ears, nose, tongue, and skin. Eyesight is the sense that allows us to see. Our ears allow us to hear sounds. Our nose lets us smell things. The sense of taste allows us to experience flavors on our tongues. And finally, our skin is sensitive to touch and temperature changes in our environment. Our eyes allow us to see. They're located on either side of our head, and they contain rods and cones that turn light into electrical signals that are sent to the brain. Our ears let us hear.

They're shaped like a funnel with two openings: one in the front and one in the back. These openings collect sound waves from outside of the body and amplify them so we can hear them better. Our noses help us smell things around us by collecting particles that have been released by objects or people whose scents we want to detect. These particles then travel through our nose where they're analyzed by cilia, tiny hairs inside our nostrils that also filter out dust particles from entering your lungs when you breathe in air through your nose (Kang et al., 2009). Our tongues help us taste food by releasing chemicals onto different areas of it when you put something into your mouth. The chemicals dissolve into liquid form once they come into contact with saliva which makes its way down into your throat; this liquid travels through a tube called a "salivary duct" where it reaches other parts of your body where nutrients need to be absorbed into cells for them to function.

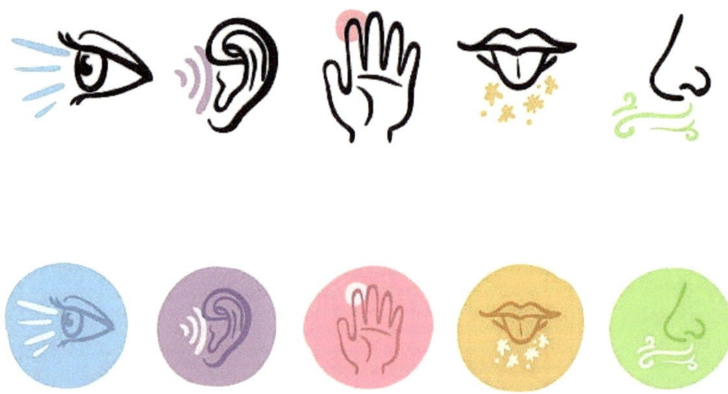

Figure 4.3: An image showing the main sensory organs.

Source: https://www.istockphoto.com/es/vector/conjunto-de-iconos-de-cinco-sentidos-gm543977278–97709643.

Inside sensation, or interoception, identifies boosts from inner organs and tissues. Numerous inside tactile and perceptual frameworks exist in people, counting the vestibular framework, the sense of adjustment detected by the internal ear and giving the recognition of spatial introduction; proprioception, body position; and nociception torment. Advance inside chemoreception- and osmoreceptors-based tangible frameworks lead to different discernments, such as starvation, thirst, suffocation, and queasiness, or distinctive automatic behaviors, such as spewing.

Sensory modality alludes to the way that data is encoded, which is comparable to the thought of transduction. The most tactile modalities can be portrayed on the premise of how each is transduced. Posting all the diverse tangible modalities, which can number as numerous as 17, includes isolating the major faculties into more particular categories, or submodalities, of the bigger sense (Jain et al., 2014). A person's tactile methodology speaks to the sensation of a particular sort of jolt. For case, the common sensation and discernment of touch, which is known as somatosensation, can be isolated into a lightweight, profound weight, vibration, tingle, torment, temperature, or hair development, whereas the common sensation and discernment of taste can be isolated into submodalities of sweet, salty, acrid, severe, fiery, and umami, all of which are based on distinctive chemicals official to tactile neurons.

Figure 4.4: An image showing sensory modality.

Source: https://www.istockphoto.com/es/foto/foto-sobre-el-tema-ar-las-palabras-ar-online-sobre-un-fondo-colorido-imagen-gm1312094423–401000800.

4.3. HUMAN SENSORY RECEPTORS

Sensory receptors are the cells or structures that recognize sensations. Jars inside the environment actuate specialized receptor cells inside the periphery on the edge system. Amid transduction, the physical jar is changed over into action potential by receptors and transmitted towards the central on edge system for planning. Unmistakable sorts of shocks are identified by unmistakable sorts of receptor cells (Igarashi et al., 2010). Receptor cells can be classified into sorts on the preface of three unmistakable criteria: cell sort, position, and work. Receptors can be classified essentially on the introduction of cell sort and their position in association to boost their sense. Receptors can help be classified essentially on the introduction of the transduction of shocks, or how the mechanical shock, light, or chemical changed the cell film potential.

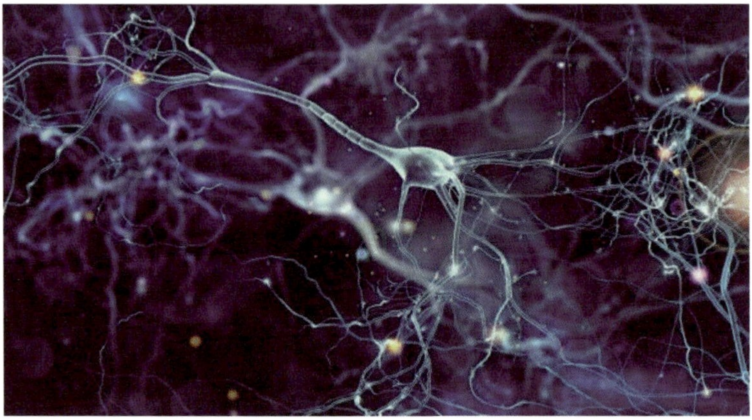

Figure 4.5: An image the human sensory receptors.

Source: https://www.istockphoto.com/es/foto/concepto-de-las-c%C3%A9lulas-de-las-neuronas-gm1053434168–281443415.

One way to classify receptors is based on their zone relative to the stimuli. An exteroceptor may be a receptor that's found near a shock of the outside environment, such as the somatosensory receptors that are found inside the skin. An interoceptor deciphers shocks from inward organs and tissues, such as the receptors that sense the increase in blood weight inside the aorta or carotid sinus.

The cells that decipher data almost the environment can be either a neuron that contains a free nerve finishing, with dendrites implanted in tissue that would get a sensation; a neuron that has a typified finishing in which the tangible nerve endings are typified in connective tissue that improves their affectability; or a specialized receptor cell, which has particular auxiliary components that decipher a particular type of boost. The torment and temperature receptors within the dermis of the skin are illustrations of neurons that have free nerve endings (Huang et al., 2022). Too found within the dermis of the skin are lamellated corpuscles, neurons with typified nerve endings that react to weight and touch. The cells within the retina that react to light boosts are a case of a specialized receptor, a photoreceptor.

A transmembrane protein receptor may be a protein within the cell layer that intervenes in a physiological alter in a neuron, most frequently through the opening of particle channels or changes within the cell signaling forms. Transmembrane receptors are actuated by chemicals called ligands. For case, an atom in nourishment can serve as a ligand for taste receptors. Other transmembrane proteins, which are not precisely called receptors, are delicate to mechanical or warm changes. Physical changes in these proteins increment ion flow over the film and can produce an activity potential or a reviewed potential within the tangible neurons.

The third classification of receptors is by how the receptor transduces shocks into film potential changes. Boosts are of three common sorts. Several boosts are particles and macromolecules that impact transmembrane receptor proteins when these chemicals diffuse over the cell layer. Some boosts are physical assortments inside the environment that impact receptor cell film conceivable outcomes (Hong et al., 2008). Other boosts consolidate the electromagnetic radiation from unmistakable light. For individuals, the only electromagnetic imperativeness that's seen by our eyes is unmistakable light.

Receptor cells can be further categorized on the premise of the sort of boosts they transduce. The diverse sorts of useful receptor cell sorts are mechanoreceptors, photoreceptors, chemoreceptors, osmoreceptors, thermoreceptors, and nociceptors. Physical jolts, such as weight and vibration, as well as the sensation of sound and body position, are deciphered through a mechanoreceptor. Photoreceptors change over light, unmistakable electromagnetic radiation, into signals. Chemical jolts can be deciphered by a chemoreceptor that translates chemical boosts, such as an object's taste or scent, whereas osmoreceptors react to a chemical solute concentrations of body liquids. Nociception translates the nearness of tissue harm, from tactile data from mechano-, chemo-, and thermoreceptors. Another physical boost that has it's possess sort of receptor is temperature, which is detected through a thermoreceptor that's either touchy to temperatures over or below normal body temperature.

4.4. SENSE ORGANS THRESHOLD

Each sense organ; eyes or nose, for an event requires an irrelevant entirety of affectation in organize to recognize a boost. This slightest entirety of boost is called the by and large edge. The through and through restrain is characterized as the slightest entirety of prompting essentially for the area of a boost 50% of the time. Through and through edge is measured by utilizing a procedure called hail area. This get-ready incorporates showing boosts of changing the drive to a subject in orchestrate to choose the level at which the subject can dependably recognize affectation in a given sense.

Differential threshold or just recognizable difference (JDS) is the humblest recognizable differentiation between two boosts or the tiniest refinement in shocks that can be judged to be particular from each other. Weber's Law is an observational law that states that the differentiate restrain may be an unfaltering division of the comparison shock. Concurring with Weber's Law, more noteworthy boosts require greater contrasts to be taken note of (Happiness et al., 2011). Human power exponents and Steven's Control Law Size estimation might be a psychophysical technique in which subjects dole out seen values of given shocks. The relationship between boost raised and perceiving heightened is depicted by Steven's control law.

Signal detection theory measures the encounter of the subject to the introduction of a boost within the presence of clamor. There's inside commotion and there's an outside commotion when it comes to flag discovery. The inside clamor begins from inactive within the anxious framework. For case, if a person with closed eyes in a dull room still sees something—a messy design of dark with intermittent brighter flashes—this is inside clamor. Outside clamor is the result of commotion within the environment that can meddle with the discovery of the jolt of intrigued. Commotion is as it were an issue in case the greatness of the clamor is expansive sufficient to meddle with the signal collection. The apprehensive framework calculates a basis, or an inside limit, for the location of a flag within the nearness of commotion (Gunaratne et al., 2019). In case a flag is judged to be over the basis, in this way the flag is separated from the commotion, and the flag is detected and seen. Mistakes in flag location can lead to untrue positives and true negatives. The sensory methods might be different based on the importance of identifying the signal. The change of the signal detection method may affect the possibility of true positives and untrue negatives.

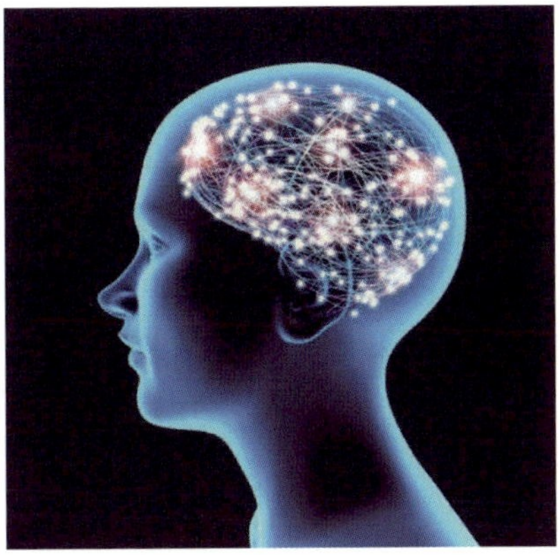

Figure 4.6: An image illustrating signal detection theory.

Source: https://www.istockphoto.com/es/foto/desarrollo-cerebral-en-ni%C3%B1os-y-adolescentes-est%C3%ADmulos-e-ideas-c%C3%B3mo-funcionan-las-gm1284452869–381586119.

Subjective visual and sound-related encounters show up to be comparable over people subjects. The same cannot be said almost taste. For illustration, there's a particle called propylthiouracil (PROP) that a few people encounter as biting, a few as nearly bland, whereas others encounter it as someplace between tasteless and sharp. There's a hereditary premise for this contrast between recognition given the same tactile jolt. This subjective contrast in taste recognition has suggestions for individuals' nourishment inclinations, and subsequently, wellbeing. When a jolt is steady and constant, perceptual tactile adjustment happens. Amid this preparation, the subject gets to be less touchy about the boost.

Organic sound-related, which is hearing, vestibular and spatial, and visual frameworks show up to break down real-world complex boosts into sine wave components, through the scientific preparation called Fourier examination. Numerous neurons have a solid inclination for certain sine recurrence components that differentiate them from others (Fiorentini et al., 2020). The way that easier sounds and pictures are encoded amid sensation can give an understanding of how recognition of real-world objects happens.

4.5. HUMAN SENSORY NEUROSCIENCE

Perception happens when nerves that lead from the tactile organs to the brain are fortified, indeed in case that incitement is irrelevant to the target flag of the tactile organ. For case, within the case of the eye, it does not matter whether light or something else fortifies the optic nerve, that incitement will come about in visual discernment, indeed in case, there was no visual boost to start with. To prove this point to yourself and in case you're a human, near your eyes; ideally in a dull room, and press gently on the exterior corner of one eye through the eyelid. You'll see a visual spot toward the interior of your visual field, close your nose.

All stimuli gotten by the receptors are transduced to a movement potential, which is carried along one or more afferent neurons towards a specific locale of the brain. Reasonable as particular nerves are committed to substantial and motors errands, particular zones of the brain are so too committed to different unmistakable and perceptual assignments. More complex planning is wrapped up over fundamental cortical districts that spread past the fundamental cortices. Each nerve, unmistakable or motor, has its claim hail transmission speed.

The reasoning of acknowledgment is concerned with the nature of perceptual inclusion and the status of perceptual data, in particular how they relate to feelings roughly, or data of, the world. Irrefutable ask into the essential disobedient of sensation and insight have driven early examiners to subscribe to diverse philosophical interpretations of insight and the judgment skills, tallying panpsychism, dualism, and authenticity (Feng et al., 2019). The lion's share of cutting-edge analysts who consider sensation and acknowledgment take on a materialistic see of the judgment skills.

4.6. THE SENSE OF SIGHT

The visual system, or sense of sight, is based on the transduction of light stuns gotten through the eyes and contributes to visual understanding. The visual framework recognizes light on photoreceptors interior the retina of each eye that produces electrical nerve driving qualities for the affirmation of moving colors and brightness. There are two sorts of photoreceptors: bars and cones. Shafts are exceptionally fragile to light but don't recognize colors. Cones recognize colors but are less unsteady to dim light. At the nuclear level, visual boosts because changes interior of the photopigment particle that leads to changes inside the layer potential of the photoreceptor cell. A single unit of light is called a photon, which is portrayed in texture science as a distribution of essentialness with properties of both a molecule and a wave.

The imperativeness of a photon is talked to by its wavelength, with each wavelength of self-evident light compared to a specific color. Even though in spite of the fact that wavelengths shorter than 380 nm drop into the shinning run. Light with a wavelength of 380 nm is blue whereas light with a wavelength of 720 nm is gloomy reddish. All other colors drop between reddish and blue at diverse centers along the wavelength scale. The three sorts of cone opsins, being tricky to assorted wavelengths of light, donate us with color vision. By comparing the development of the three different cones, the brain can remove color information from visual boosts. For outline, a shining blue light that consolidates a wavelength of generally 450 nm would activate the "rosy" cones irrelevantly, the "green" cones barely, and the "blue" cones overwhelmingly (Elsamani et al., 2014). The relative incitation of the three different cones is calculated by the brain, which sees the color as blue.

Figure 4.7: An image showing an eye which represents the sense of sight.

Source: https://www.istockphoto.com/es/foto/ojo-xxl-gm108220711–7593226.

However, cones cannot respond to low-intensity light, and bars don't sense the color of light. In this manner, our low-light vision is, in essence, in grayscale. In other words, in a dull room, everything shows up as a shade of gray. If you think simply can see colors within the dim, it is most likely since your brain knows what color something is and is depending on that memory. There are a few differences as to whether the visual framework comprises one, two, or three submodalities. Neuroanatomists for the most part respect it as two submodalities, given that diverse receptors are capable of the discernment of color and brightness. A few contend that stereopsis, the recognition of profundity utilizing both eyes, moreover constitutes a sense, but it is by and large respected as a cognitive work of the visual cortex of the brain where designs and objects in pictures are recognized and deciphered based on already learned data. Usually called visual memory.

The failure to see is called visual impairment. Visual impairment may result from harm to the eyeball, particularly to the retina, harm to the optic nerve that interfaces each eye to the brain, and stroke. Brief or changeless visual deficiency can be caused by the harms of medicines. Individuals who are dazzled from corruption or harm to the visual cortex, but still have useful eyes, are competent in a few levels of vision and response to visual jolts but not a cognizant recognition; this can be known as blindsight. Individuals with blindsight are as a rule not mindful that they are responding to visual sources, and instep fair unknowingly adjusts their behavior to the boost (Elina et al., 2016). On February 14, 2013, analysts created a neural embed that gives rats the ability to sense infrared light which for the primary time gives living animals with unused capacities, rather than supplanting or increasing existing capacities.

Figure 4.8: An image showing a visually impaired individual.

Source: https://www.istockphoto.com/es/foto/estudiante-ciego-en-la-universidad-gm1202384535–345177311.

Concurring to Gestalt Brain research, people see the total of something without a doubt on the off chance that it isn't there. The Gestalt Law of Organization states that people have seven factors that offer help to bunch what is seen into plans or bunches: Common Fate, Resemblance, Closeness, Closure, Symmetry, Coherence, and Past Inclusion. The Law of Common fate says that objects are driven along the smoothest way. People take after the incline of development as the lines/dots stream. The Law of Closeness insinuates the gathering of pictures or objects that are comparative to each other in a couple of points. This can be due to shade, color, degree, shape, or other qualities you'll recognize (Ebuehi & Oyewole, 2007). The Law of Region states that our minds like to bunch based on how close objects are to each other. We may see 42 objects in an assembly, but able to as well see three bunches of two lines with seven objects in each line.

The Law of Closure is the thought that we as individuals still see a full picture in fact within the occasion that there are gaps interior that picture. There may be holes or parts misplaced from a section of a shape, but we would still see the shape as a whole. The Law of Symmetry insinuates a person's slant to see symmetry around a central point. A case would be when we utilize walled-in areas in composing. We tend to see all of the words inside the enclosures as one section instead of individual words inside the parentheses. The Law of Movement tells us that objects are assembled by their components and after that seen as an aggregate. This more frequently than not happens when we see covering objects. We'll see the covering objects with no interruptions. The Law of Past Association implies the penchant for individuals got to categorize objects concurring to past experiences under certain circumstances. In case two objects are customarily seen together or interior close closeness to each other the Law of Past Experience is usually seen.

4.7. THE SENSE OF HEARING

Hearing is the transduction of sound waves into a neural salute that's made conceivable by the structures of the ear. The wide, full structure on the sidelong point of see of the head is known as the auricle. After the sound-related canal is the tympanic film, or ear drum, which vibrates after it is struck by sound waves. The auricle, ear canal, and tympanic film are routinely inferred to as the outside ear. The center ear comprises a space crossed by three little bones called the ossicles. The three ossicles are the malleus, incus, and stapes, which are Latin names that are by and expansive deciphered as a pound, press square, and stirrup (Difonzo et al., 2019). The malleus is related to the tympanic film and communicates with the incus. The incus, in turn, verbalizes with the stapes. The stapes are at that point related to the interior ear, where the sound waves will be transduced into a neural accost.

Figure 4.9: An image showing an eye which represents the sense of hearing.

Source: https://www.istockphoto.com/es/foto/oreja-y-cierre-de-las-manos-femeninas-copiar-espacio-papel-desgarrado-fondo-amarillo-gm1170664224–324025829.

The center ear is related to the pharynx through the Eustachian tube, which makes a distinction between equilibrate talk about weight over the tympanic layer. The tube is commonly closed but will pop open when the muscles of the pharynx contract in the midst of swallowing or yawning. Mechanoreceptors turn development into electrical nerve beats, which are found inside the internal ear. Since sound is vibration, multiplying through a medium such as examination, the discovery of these vibrations, that's the sense of the hearing, maybe a mechanical sense since these vibrations are mechanically conducted from the eardrum through a course of action of minor bones to hair-like strands inside the internal ear, which distinguish mechanical development of the strands interior a expand of around 20 to 20,000 hertz, with critical assortment between individuals (Abebe et al., 2006). Hearing at tall frequencies diminishes with an increase in age. Disappointment to tune in is called deafness or hearing impedance.

Sound can be identified as vibrations conducted through the body by a tactician. Lower frequencies that can be listened to are identified this way. A few hard-of-hearing individuals can decide the heading and area of vibrations picked up through the feet. Studies relating to try out began to extend in number towards the last mentioned conclusion of the nineteenth century. During this time, numerous research facilities within the Joined together States started to make new models, charts, and disobedient that all related to the ear. There could be a department of Cognitive Brain research committed entirely to try out. They call it Sound-related Cognitive Brain research. The most point is to get why people can utilize sound in considering the exterior of really saying it. Relating to Sound-related Cognitive Brain research is Psychoacoustics. Psychoacoustics is more pointed to individuals inquisitive about music. Haptics, a word utilized to allude to both taction and kinesthesia, has numerous parallels with psychoacoustics.

4.8. THE SENSE OF TOUCH

The sense of touch is said to be part of the somatosensory system. Somatosensation is the group of tactile modalities that are related to touch and interoception. The modalities of somatosensation incorporate weight, vibration, light touch, tickle, tingle, temperature, torment, and kinesthesia. Somatosensation moreover called tactician could be a discernment resulting from the enactment of neural receptors, for the most part within the skin counting hair follicles, but within the tongue, throat, and mucosa. An assortment of weight receptors react to varieties in weight; firm, brushing, and maintained, among others (De Kock & Magano, 2020). The touch sense of tingling caused by creepy crawly nibbles or sensitivities includes uncommon itch-specific neurons within the skin and spinal cord. The misfortune or impedance of the capacity to feel anything touched is called material anesthesia. Paresthesia may be a sensation of shivering, pricking, or deadness of the skin which will result from nerve harm and may be lasting or brief.

Figure 4.10: An image showing a hand which represents the sense of touch.

Source: https://www.istockphoto.com/es/foto/una-mano-toca-un-musgo-verde-del-bosque-punto-de-vista-luz-solar-gm1249853647–364366754.

Two sorts of somatosensory signals that are transduced by free nerve endings are pain and temperature. These two modalities utilize thermoreceptors and nociceptors to transduce temperature and torment boosts, individually. Temperature receptors are invigorated when neighborhood temperatures contrast with body temperature. A few thermoreceptors are delicate to fair cold and others to fair warm. Nociception is the sensation of possibly harming jolts. Mechanical, chemical, or warm jolts past a set edge will evoke difficult sensations. Focused or harmed tissues discharge chemicals that actuate receptor proteins within the nociceptors.

Low recurrence vibrations are detected by mechanoreceptors called Merkel cells, moreover known as sort I cutaneous mechanoreceptors. Merkel cells are found within the stratum basale of the epidermis. Profound weight and vibration are transduced by lamellated corpuscles, which are receptors with encapsulated endings found profound within the dermis, or subcutaneous tissue. A light touch is transduced by the typified endings known as material corpuscles. Follicles are moreover wrapped in a plexus of nerve endings known as the hair follicle plexus (Daher et al., 2020). These nerve endings identify the development of hair at the surface of the skin, such as when a creepy crawly may be strolling along the skin. Extending of the skin is transduced by extended receptors known as bulbous corpuscles. Bulbous corpuscles are moreover known as Ruffini corpuscles or sort II cutaneous mechanoreceptors.

The heat receptors react to infrared radiation; hence the radiation can occur in specialized hearing organs, for instance, in pit vipers. The heat receptors found on human skin are very different from the homeostatic thermoreceptors in the brain, (that is, the hypothalamus), which give feedback on internal body temperature.

4.9. THE SENSE OF TASTE

The gustatory framework or the sense of taste is the tangible framework that's mostly mindful of the discernment of taste. Several recognized submodalities exist inside taste: sweet, salty, acrid, biting, and umami. Exceptionally later investigation has proposed that there may be a 6th taste submodality for fats or lipids. The sense of taste is regularly confounded with the recognition of flavor, which comes about from the multimodal integration of gustatory and olfactory sensations. Philippe Mercier – The Sense of Taste – Google Craftsmanship Project Within the structure of the lingual papillae are taste buds that contain specialized gustatory receptor cells for the transduction of taste boosts (Chen & Opara, 2013). These receptor cells are delicate to the chemicals contained inside nourishments that are ingested, and they discharge neurotransmitters based on the sum of the chemical within the nourishment. Neurotransmitters from the gustatory cells can activate tactile neurons within the facial, glossopharyngeal, and vagus cranial nerves. Salty and acrid taste submodalities are activated by the cations Na+ and H+, individually. The other taste modalities result from nourishment atoms official to a G protein-coupled receptor. A G protein flag transduction framework eventually leads to depolarization of the gustatory cell. The sweet taste is the affectability of gustatory cells to the presence of glucose or sugar substitutes broken down within the spit.

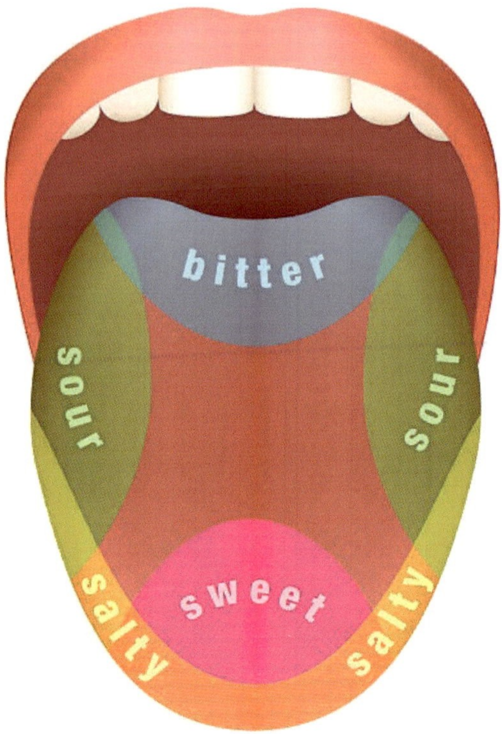

Figure 4.11: An image showing the tongue which represents the sense of taste.

Source:https://www.istockphoto.com/es/vector/lengua-sabor-%C3%A1reas-gm524886192–92277189.

The bitter taste is comparative too sweet in that nourishment atoms tie to G protein-coupled receptors. The taste known as umami is regularly alluded to as the savory taste. Like sweet and sharp, it is based on the actuation of G protein-coupled receptors by a particular molecule. Once the gustatory cells are actuated by the taste particles, they discharge neurotransmitters onto the dendrites of tangible neurons. These neurons are a portion of the facial and glossopharyngeal cranial nerves, as well as a component inside the vagus nerve devoted to the choking reflex. The facial nerve interfaces to taste buds within the front third of the tongue. The glossopharyngeal nerve connects to taste buds within the back two-thirds of the tongue (Chowdhury & Das, 2015). The vagus nerve interfaces to taste buds within the extraordinary back of the tongue, skirting on the pharynx, which is touchier to harmful jolts such as severity.

The flavor depends on odor, surface, and temperature as well as on taste. People get tastes through tangible organs called taste buds, or gustatory calyculi, concentrated on the upper surface of the tongue. Other tastes such as calcium and free greasy acids may moreover be essential tastes but have however to gotten far-reaching acknowledgment. The failure to taste is called ageusia. There may be an uncommon marvel when it comes to the gustatory sense. It is called Lexical-Gustatory Synesthesia. Lexical-Gustatory Synesthesia is when individuals can "taste" words. They have detailed flavor sensations they aren't eating. When they perused words, listen to words, or indeed envision words. They have detailed not as it were basic flavors, but surfaces, complex flavors, and temperatures as well.

4.10. THE SENSE OF SMELL

Just like the sense of taste, the sense of smell, or the olfactory framework, is additionally responsive to chemical stimuli. Not at all like taste, there are hundreds of olfactory receptors, each authoritative to a specific atomic include. Odor atoms have an assortment of highlights and, in this way, energize particular receptors more or less emphatically. This combination of excitatory signals from diverse receptors makes up what people see as the molecule's smell. The olfactory receptor neurons are found in a little locale inside the prevalent nasal depression (Chira et al., 2020). This locale is alluded to as the olfactory epithelium and contains bipolar tactile neurons. Each olfactory tangible neuron has dendrites that expand from the apical surface of the epithelium into the bodily fluid lining the depth.

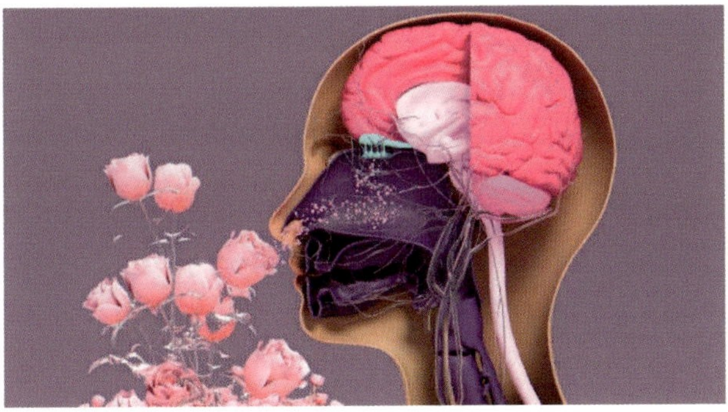

Figure 4.12: An image illustrating the sense of smell.

Source: https://www.istockphoto.com/es/foto/sistema-olfativo-sistema-sensorial-utilizado-para-oler-sentidos-de-olfato-componentes-gm1302713332–394373671.

As airborne particles are breathed in through the nose, they pass over the olfactory epithelial locale and break down into bodily fluid. These odorant particles tie to proteins that keep them broken down within the bodily fluid and offer assistance to transport them to the olfactory dendrites. The odorant–protein complex ties to a receptor protein inside the cell film of an olfactory dendrite. These receptors are G protein–coupled, and will create a graded layer potential within the olfactory neurons. In the brain, olfaction is processed by the olfactory cortex. Olfactory receptor neurons within the nose vary from most other neurons in that they kick the bucket and recover on a regular premise. The failure to scent is called anosmia. A few neurons within the nose are specialized to identify pheromones.

Misfortune of the sense of scent can result in nourishment tasting insipid. An individual with an impeded sense of scent may require extra zest and flavoring levels for nourishment to be tasted. Anosmia may too be related to a few introductions of gentle sadness, since the misfortune of delight of nourishment may lead to a common sense of losing hope. The capacity of olfactory neurons to supplant themselves diminishes with age, driving to age-related anosmia (Coggins, 2007). This clarifies why a few elderly individuals salt their nourishment more than more youthful individuals do. Causes of Olfactory brokenness can be caused by age, presentation to poisonous chemicals, viral diseases, epilepsy, a few sort of neurodegenerative infection, head injury, or as a result of another clutter.

As studies in olfaction have proceeded, there has been a positive relationship between its brokenness or degeneration and early sign of Alzheimer's and intermittent Parkinson's malady. Numerous patients do not take note of the decrease in scent sometime recently being tried. In Parkinson's Malady and Alzheimer's and the olfactory shortage is displayed in 85 to 90% of the early onset cases. There is proof that the decrease of this sense can go before the Alzheimers or Parkinson's Illness for a long time. In spite of the fact that the shortage is displayed in these two maladies, as well as others, it is imperative to form note that the seriousness or size changes with every disease. This has brought to light a few recommendations that olfactory testing may be utilized in a few cases to help in separating numerous neurodegenerative illnesses.

Those who were born without a sense of scent or have a harmed sense of scent as a rule complain approximately 1, or more, of 3 things. Our olfactory sense is additionally utilized as a caution against awful nourishment. In case the sense of scent is harmed or not there, it can lead to individual contracting nourishment harming more regularly. Not having a sense of scent can moreover lead to harmed connections or frailties inside the connections since of the failure of the individual to not scent body odor (Chen et al., 2022). Finally, scent impacts how nourishment and drink taste. When the olfactory sense is harmed, the fulfillment from eating and drinking isn't as conspicuous.

4.11. OTHER HUMAN SENSES

An inner sensation and recognition too known as interoception is any sense that's ordinarily fortified from inside the body. These include various tactile receptors in inner organs. Interoception is thought to be atypical in clinical conditions such as alexithymia. A few cases of particular receptors are: Hunger is administered by a set of brain structures; for example, the hypothalamus, which is mindful of vitality homeostasis. Pulmonary extend receptors are found within the lungs and control the respiratory rate. Peripheral chemoreceptors within the brain screen the carbon dioxide and oxygen levels within the brain to donate a recognition of suffocation if carbon dioxide levels get as well tall.

The chemoreceptor trigger zone is an extension of the medulla inside the brain that gets inputs from blood-borne drugs or hormones and communicates with the retching center. Chemoreceptors inside the circulatory system in addition degree salt levels and affect thirst on the off chance that they get as well tall; they can as well respond to tall blood sugar levels in diabetics. Cutaneous receptors inside the skin not because it reacted to touch, weight, temperature, and vibration, but moreover respond to vasodilation inside the skin such as reddening. Extend receptors inside the gastrointestinal tract sense gas distension which can result in colic torment. Incitement of unmistakable receptors inside the esophagus results in sensations felt inside the throat when swallowing, regurgitating, or amid destructive reflux (Castro et al., 2013). Tangible receptors in pharynx mucosa, comparable to touch receptors inside the skin, sense farther objects such as mucous and food that will result in a choking reflex and comparing choking sensation.

Incitement of tactile receptors within the urinary bladder and rectum may result in discernments of fullness. Stimulation of extended sensors that sense the widening of different blood vessels may result in torment, for illustration migraine caused by vasodilation of brain arteries. Cardioception alludes to the recognition of the action of the heart. Opsins coordinate DNA harm in melanocytes and keratinocytes can sense bright radiation, which plays a part in pigmentation and sunburn. Baroreceptors transfer blood weight data to the brain and keep up legitimate homeostatic blood pressure. The perception of time is additionally in some cases called a sense, even though not tied to a particular receptor.

Tense is a type of human sense, or an ability that humans possess. The sense of tense is what allows people to distinguish between past, present, and future. While animals may have the ability to experience time in this way, they do not have tense per se: they do not experience the world as being divided into three distinct categories. Humans can experience time in many ways—as duration (how long something lasts), as succession (events happening one after another), as change over time, or as anticipation (what will happen next). This has led some researchers to believe that our sense of tension helps us plan for the future (Bodyfelt et al., 2008). For example, when you hear a sound then you know that it happened in the past because it has already finished – but if you see someone walking then you know that it happened in the present because it is still happening now – but if you feel something cold on your skin then you know that it is going to happen in the future because it hasn't happened yet.

Figure 4.13: An image illustrating the sense of tense.

Source: https://www.istockphoto.com/es/vector/concepto-de-cambio-de-humor-muchas-emociones-rodean-a-las-mujeres-j%C3%B3venes-con-gm1286189024–382774459.

The sense of spence is a human sense. It is one of the five senses in humans and it enables us to feel pressure and vibration. The sense of spence can be used to distinguish between different pressures, such as hard or soft, and it allows us to feel vibrations through the skin.

4.12. CHALLENGES FACING HUMAN SENSES

The human senses are a powerful tool for gathering information about our world. They help us navigate the physical and social environments around us, and they are integral to our ability to function as human beings. However, at the same time, these very same senses can sometimes be a source of frustration—when a smell or sound is unpleasant or causes you discomfort, for example (Birol et al., 2015). While some of these challenges can be solved with technology (such as noise-canceling headphones or an air purifier), others will require conscious effort from both individuals and society at large.

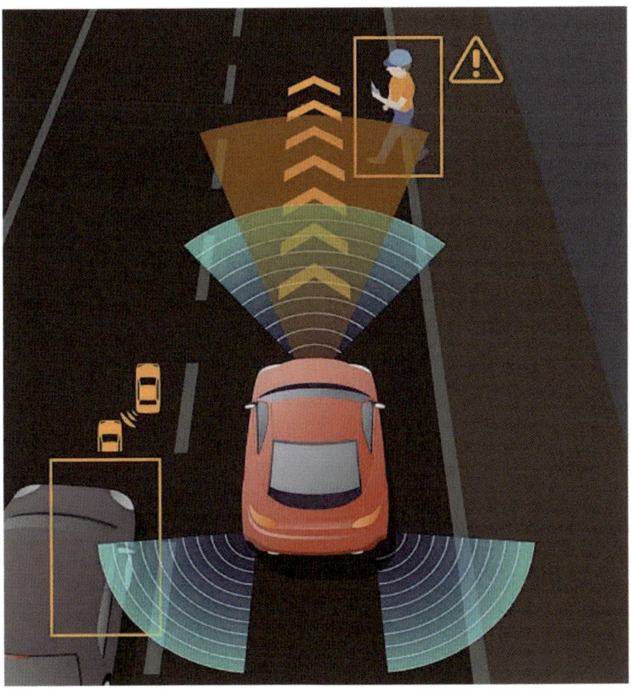

Figure 4.14: An image illustrating current technology as a challenge to human senses.

Source: https://www.istockphoto.com/es/vector/sistema-avanzado-de-asistente-de-conducci%C3%B3n-monitoreo-de-punto-ciego-gm616893138–107194623.

Each sense has its own challenges that affect its ability to work properly when they are damaged or impaired by disease or injury. For example: if someone loses their sight due to injury or illness then they will have trouble navigating around their environment because they cannot see where they are going; if someone loses their hearing then they will find it difficult making friends because they cannot communicate effectively with others; if someone loses their sense of smell then they may not be able to tell if food has gone bad which could lead them into eating something that makes them sick; if someone loses their sense.

The following are just some of the challenges facing each sense: *Touch:* We live in a world that is increasingly digital. Touch has been replaced by clicks and taps on screens, which can often feel impersonal and less personal than direct human contact. *Sight:* As more people spend time online, vision becomes distorted by advertisements that pop up on your screen every time you visit a site or app. This can cause eyestrain, headaches and other vision issues over time (Biró et al., 2020). *Hearing:* Noise pollution is becoming an increasingly serious problem in cities around the world—in addition to being an annoyance, it can cause hearing loss over time in those who work in noisy environments regularly. While our sense of smell can be affected by many factors including age, health conditions and medications, it's important to understand how aging affects your ability to smell so you can take measures to protect it as you get older.

4.13. SOLUTIONS TO THE CHALLENGES FACING HUMAN SENSES

Human senses are crucial to our survival. Our ability to hear, see, smell and taste all allow us to interact with our environment in order to survive. With the increasing reliance on technology and its effects on human interaction, these senses are being ignored and left behind. We need to find ways to make sure that humans continue to connect with their surroundings (Anton & Luciano, 2007). One way we can do this is by creating solutions for each sense that help build connections between people and the world around them. This will ensure that the human race continues to thrive.

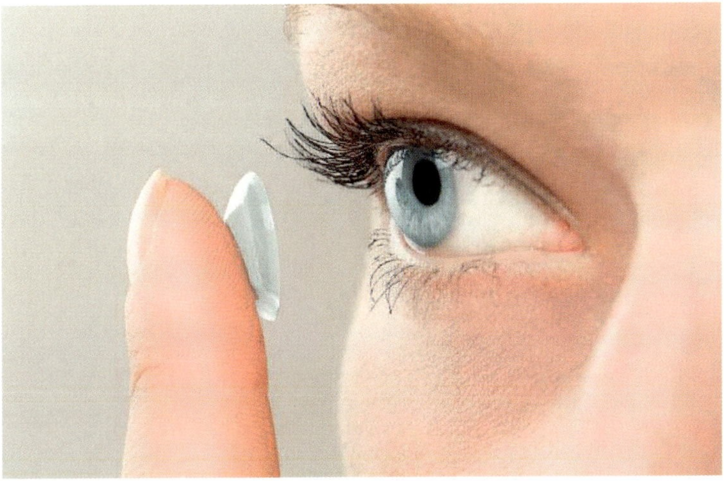

Figure 4.15: An image showing the use of contact lenses as a solution to challenges in the sense of sight.

Source: https://www.istockphoto.com/es/foto/mujer-inserci%C3%B3n-de-una-lente-de-contacto-en-los-ojos-gm507172508–84593281.

For instance, the sense of sight is the most important, as it is the primary way we interact with our environment. This can cause issues in certain circumstances, such as when a person has a visual impairment or an inability to see at all. This challenge is relatively easy to solve with glasses and contact lenses (Beekman et al., 2022). However, there are other senses that are just as important and are more challenging to overcome.

CHAPTER 5

OLFACTION

CONTENTS

5.1. Introduction 108

5.2. The Psychology and Physiology of Athletic Performance in Humans 114

5.3. Theories of Olfaction 119

5.4. Parameters of Olfactory Stimulation 120

5.5. Olfaction-Related Physiological Mechanisms 120

5.6. Olfactory Quality Classification Systems 121

5.7. Innovations in the Sensory Reception of Odor 123

5.8. Investigating Different Conceptions of Radiation 124

5.9. The Shift Theory, Developed By Raman 125

5.10. Theories Regarding Infrared Spectroscopy 126

5.11. Theories on the Pattern of Stimulus 126

5.12. Chemistry Theory 127

5.1. INTRODUCTION

The information that the senses provide is the foundation upon which the brain constructs its view of the external world. One of the five senses that is frequently misunderstood is olfaction. This is due, in part, to the fact that the hippocampus plays a significant role in the processing of olfactory information. Despite this, there is no denying the influence that smell has on behavior and cognition (Anton & Luciano, 2007). Multiple studies have found that there is a significant connection between the way people experience scent and the reactions it elicits in them. There are many facets of life in which olfactory cues play an important role, including dating, parenting, and feeding. The sense of smell may also play a role in cognitive processes, the regulation of mood, and emotional responses. Failure of the olfactory system is directly linked to the development of severe disorders such as neuropsychiatric depression and difficulty with eating.

As a consequence of this, the link between the two conditions has been established. Because of this, having a comprehensive knowledge of the molecular processes that are involved in olfaction is essential for comprehending animal behavior in both normal and abnormal circumstances. Olfactory sensory neuroepithelial cells are remarkable in that they send direct projections to the main olfactory bulb without the need for an intermediary brain region. These cells express odor receptors that are only found in the nasal cavity. The Molecular Organization of the Brain (MOB) is comprised of numerous cell types to process olfactory information and transmit it to higher brain regions. Sanchez -research Gonzalez investigated the geographical origins, heterogeneity, and genetic composition of the olfactory bulb interneurons to gain a better understanding of how they develop from various types of progenitor cells located throughout the brain. In his contribution to the research on the cytoarchitecture of the olfactory circuit, Fengyi Liang analyzes the significance of the cellular interaction that occurs between olfactory receptor neurons and olfactory sustentacular cells.

Multiple olfactory processes are dependent on intricate cell-to-cell interactions, which are also controlled by the neuromodulatory systems. One of these processes is the ability to recognize odors. The endocannabinoid system is being investigated by Terral as a potential link between olfactory information and behavioral responses. Olfactory structures serve as the focal point for peripheral signals that regulate eating behavior by monitoring the nutritional state of the organism. This is accomplished through the monitoring of odors. Structures of the sense of smell: As demonstrated by Wu, a negative energy balance is shown to affect the mitral cell activity found in the MOB. It is interesting to note that these changes have been linked to a reduction in the sensitivity of the sense of smell (Yağcı & Göğüş, 2008).

The universal rules of information processing in the brain that are essential to the survival of species can be studied using olfactory circuits as a model system. This can be done because the ability to smell is essential to all living things. Multiple studies have demonstrated that olfactory cues can accurately predict mate preference as well as behavior that is sexually motivated.

According to the findings of Fraichard's research, the process of choosing a mate involves an enzyme that breaks down odors. To give an example, they demonstrate that the enzyme UDP-glycosyltransferase, which is found in Drosophila sensory neurons, is necessary for the detection of sex pheromones. In addition to this, Liu presents an in-depth look at the genetics and evolution of chemosensory detection, focusing on its potential role in the regulation of physiological processes such as the detection of pheromones. According to the findings of recent studies, chemosensitivity is an essential component of the universal mechanism that all cells employ to communicate with their surroundings. Chemosensory cues such as smells have the potential to have global repercussions on a variety of levels, ranging from fundamental survival mechanisms to the state the economy is in right now. Spraying with the insect *Diachasmimorpha longicaudata*, also known as the Ashmead wasp, is one method of preventing fruit flies.

After researching the transcriptomes of parasitoid wasps, Tang discovered that these wasps' olfactory genes become active in response to the odors given off by their hosts. The question of whether or not males and females have the same level of olfactory sensitivity was investigated by Wang and his colleagues using a variety of analytical techniques. According to a wide variety of studies, sex-specific behavior can be influenced by extrinsic cues such as courting melodies, colors, and chemosensory signals (Yang & Boyle, 2016). Communication between the brain and the environment can also play a role. The fact that chemosensory receptor genes in zebrafish are more strongly expressed in males than in females hints at the existence of brain circuitry that is specific to each gender. Tasmin L. Rymer conducted a review of previous research on the function of smell in mouse paternal behavior. She highlights the significance of 10 genes linked to aggressive behaviors toward intruders and the identification of pups. Finally, this Special Issue focuses on the most recent developments in the field of olfactory research, which paves the way for new avenues of investigation across a wide range of academic fields, from genetics to behavior.

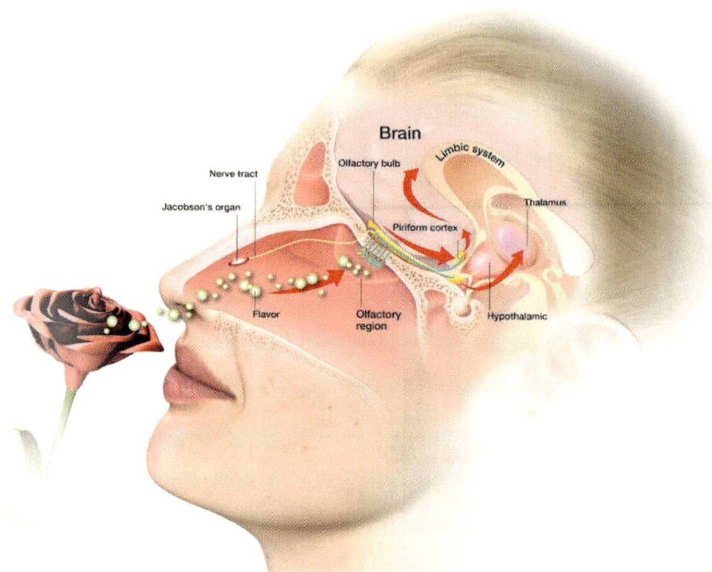

Figure 5.1: Olfactory.

Source: https://www.richardsonthebrain.com/olfactory-pathways.

The sense of smell has traditionally been regarded as less important than the other senses. Several factors contribute to this, including the difficulty of accessing anatomical structures, a lack of theoretical understanding of the nature of the physical stimulus, and a general misconception about the significance of the human olfactory sense. Each of these factors plays a role in causing this. There are still issues with some of these characteristics, but new potential solutions are being investigated. Throughout the past few years, olfactory stimulation theory has not made any significant headway.

The concept that the sense of smell is unimportant is being challenged by an expanding number of researchers in the scientific community. Because of this direct contact with airborne pollutants, the study of olfaction from a toxicological or environmental point of view is especially compelling. Because these olfactory receptors appear to function in a certain manner, airborne pollutants and hazardous substances have the potential to cause damage to a person. Because odor receptors can be affected both directly and indirectly by blood-borne toxins, they are the most sensitive brain tissue when it comes to airborne contaminants. Blood-borne toxins can affect odor receptors in one of two ways. In this article, olfactory systems and how the sense of smell influences human behavior are both investigated (Warnakulasooriya et al., 2020). Due to the limited amount of space, we can only describe the findings in a general sense. It has been determined which primary sources are essential for further research. Anatomy and physiology of the organs located in the periphery tubes within the bronchial tree.

When we breathe normally, only about 2% of the air we take in reaches the olfactory epithelium. This is because the olfactory epithelium is located deep within the nose. As it passes through the nasal epithelium, the air that we breathe in experiences changes in both its temperature and its relative humidity. The material is deposited by the cilia on the nasal mucosa, and then it is swept in the direction of the throat. Mucous is produced in the nasal epithelium by goblet cells and glands located in the mucosa. It is believed that the hypothalamus, communicating with the vascular tissue via the vidian nerve, is responsible for the rhythmic pattern of mucosal veins and cavernous tissue contracting and expanding. One of the nasal cavities will have a low flow resistance and moist nasal mucosa during the peak of the cycle, while the other will have a high flow resistance and dry nasal mucosa. The olfactory epithelium is an epithelium that can be found in the nasal cavity. It is responsible for transmitting smells.

The length of the human olfactory epithelium ranges from approximately two to four centimeters, and it contains approximately ten thousand receptor cells. This region can be found at the very tip of the nasal canals. Bowman's glands are responsible for the production of mucus, which acts as a protective layer. The olfactory epithelium is made up of three different types of cells: receptor cells, basal cells, and sustentacular cells. The receptors are similar to bipolar retinal cells both in terms of their function and their structure. Their stationary cilia can be found at a depth of approximately 160 pRm within the mucus. It is believed that receptor sites can be found in the cilia. Unmyelinated axons of receptors travel through the cribriform plate in the olfactory bulb to connect and form synapses. A rudimentary topographic map serves as a connection between the olfactory epithelium and the olfactory bulb.

To provide sustenance to the receptor cells, the sustentacular cells secrete an unknown chemical that no one is sure what it is or what it consists of.

In addition, basal cells are present in the olfactory epithelium; however, the function of these cells is not well understood.

- *Trigeminal Innervation.* The non-olfactory portion of the nasal cavity is supplied with free terminals by the ethmoid branch of the trigeminal nerve. The pharyngeal epithelium may contain some of these terminals as well (Tsai et al., 2021). The synapses that are associated with the trigeminal nerve are housed in the pons trigeminal nucleus. Olfactory structures that are positioned in the center Shepherd has studied the central nervous system's olfactory sense's connections and ultrastructure. In most other types of examinations, including background information is considered to be standard practice.

Syndrome of the Irritable Bowel Axons of receptor cells connects within the glomeruli of mitral and periglomerular cells. There are 25,000 receptor axons in each glomerulus, and each glomerulus is connected to 25 mitral cells. Axons originating from mitral cells are what make up the lateral olfactory tract, which is responsible for transmitting impulses to the brain. Mitral cells in the bulb are equipped with axon collaterals that allow them to communicate with other cells found within the bulb. Periglomerular cells, which are approximately 20 times more numerous than mitral cells, are responsible for the horizontal connection of the glomeruli.

- *Periglomerular cells.* Granular cells in the deepest levels of the olfactory bulb form dendrodendritic connections and axodendritic connections with mitral cells and each other, as well as with centrifugal neurons. These connections are found in the olfactory bulb (Tuorila & Monteleone, 2009). Centrifugal neurons typically form synapses with granule cells located in the contralateral and ipsilateral anterior olfactory nucleus as well as the ipsilateral diagonal band of Broca. In later stages of development, synapses are formed between the centrifugal fibers and the periglomerular cells.

An area of the brain called the olfactory cortex is responsible for smelling things. It is not uncommon for fibers to branch out from the central olfactory tract in the amygdaloid nucleus of the cortex, the prepyriform cortex, and the nucleus of the lateral olfactory tract. These axons don't travel very far after they leave the pyramidal cells in the outer molecular layer of the cortex, where they form axodendritic synapses. Mitral cells in the lateral olfactory tract send axons to pyramidal cells in this layer. Axons are responsible for carrying secondary fibers away from the olfactory cortex and into other parts of the brain. This nucleus is also responsible for the transmission of fibers to the granule cells located on the opposite side of the olfactory bulb, as well as to the medial forebrain bundle, which terminates in the hypothalamus.

It is well established that prepyriform cortical fibers can be found in the amygdala, the hypothalamus, and most likely the hippocampus as well. The amygdala is the source of axons that are sent to the prepyriform cortex, the hippocampus, and the hypothalamus. The following list of secondary fiber connections is by no means an exhaustive compilation of all possible combinations. Only the most vital links have been included here so that we can keep things as straightforward as possible. Recent research has uncovered a greater number of synaptic connections between the central nervous system (CNS) and the olfactory system (Short et al., 2021). These connections travel through the orbitofrontal cortex on their way to and from the thalamus, starting from the amygdala and the prepyriform cortex. Neocortical projections are found in this region, as opposed to the more typical allocortical and limbic projections.

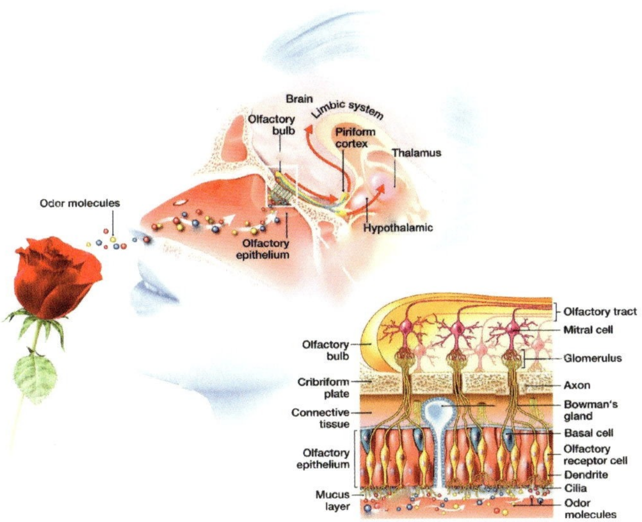

Figure 5.2: 5+ Olfactory.

Source: https://thumbs.dreamstime.com/b/olfactory-sense-sense-of-smell-detailed-illustration-of-the-olfactory-region-medically-illustration-206400971.jpg.

5.2. THE PSYCHOLOGY AND PHYSIOLOGY OF ATHLETIC PERFORMANCE IN HUMANS

Molecules floating through the air with molecular weights ranging from 15 to 300 can stimulate the olfactory system. The intensity of the stimuli is proportional to the number of odorous molecules that are brought into contact with the olfactory epithelium. The rate at which the odorant's intensity is perceived to increase as its concentration grows is a logarithmic function of concentration. This rate is influenced by the water solubility of the odorant as well as the chemical functional groups that it contains. It is more difficult to investigate threshold because it is commonly defined as the stimulus concentration that a person can perceive at least 50% of the time or less (Ayzeli et al., 2020). This makes it more difficult to determine the threshold. When exposed to the same input for an extended period, one's senses may begin to change in response, a phenomenon known as sensory adaptation. It is recommended that stimuli be given sequentially rather than sequentially to prevent cross-adaptation. There are many different approaches, pieces of apparatus, and individual differences that can be used to measure thresholds.

The categorization of odors according to their quality has been the subject of debate for millennia. There have been several hypotheses proposed as potential explanations for olfactory quality. Beets and Amoore put forward a stereo chemical olfactory hypothesis, which they based on the locations of receptors on cilia. These specific locations can make room for molecules that have the same three-dimensional shape. The scientific literature on organic chemistry served as a source of inspiration for their primary fragrances from an esthetic point of view. Because the relationships between reported odor qualities, receptor locations, and key smells aren't perfect, and because there are notable exceptions to the norms, it's difficult to draw firm conclusions about these topics. According to Davies' olfactory puncture and penetration theory, olfactory chemicals can depolarize receptor cells. It was hypothesized that some receptor membranes were more resistant to being punctured than others, which could affect the quality of the odor. There has been no evidence found to support either the existence of a specific rate of diffusion or the presence of odorant molecules at receptors.

To explain the characteristics of odors, Wright put forward a molecular vibrational theory. Although the vibrational frequency of molecules in the far-infrared is used to determine the quality of an odor, the potency of an odor is determined by factors such as its volatility, adsorption, and the degree to which it is aqueous lipid-soluble (Sutrisna et al., 2017). This concept has been met with a significant amount of skepticism. The idea of an olfactory model that incorporates both space and time has been put forward. The olfactory quality of this chromatographic analog is affected by how odorant molecules are distributed across the olfactory epithelium as a whole. This new viewpoint is only supported by a limited amount of evidence.

In conclusion, even though there are many theories concerning olfaction, the vast majority of them do not appear to be supported by the findings of scientific research. Although the majority of these theories concern the structure of the molecule, it is not known what aspect of a molecule's physical makeup is responsible for its odor. The problem is made more complicated by the fact that the mechanism behind the receptors is unknown. The Olfactory Bulb of the Central Nervous System is responsible for the Olfactory System. Because of its structure, the olfactory bulb is more than just a simple "telephone repeater" station. This is demonstrated by its anatomy. At the level of the glomerulus, sensory cells excrete their contents into synaptic neuropil which is made up of the dendrites of mitral and periglomerular cells. It appears that the fundamental function of mitral cells, which are the primary neurons in the olfactory bulb, is the transfer of sensory information. Despite this, the network of periglomerular cells makes it possible for the glomerulus to carry out a significant amount of signal processing.

At the glomerular level, there appears to be odor selectivity, and the spatial and temporal resolution of the input data is strengthened by inhibitory synaptic processes. Synapses formed between mitral cells, granule cells, and the axons of centrifugal cells are responsible for the formation of these linkages. The periglomerular cells are just as susceptible to being damaged by centrifugal force. It has been demonstrated that the synaptic connections in the inner layers of the olfactory bulb create recurrent inhibitory loops. These loops are responsible for the production of the oscillatory electroencephalogram, which can be monitored using macro electrodes in the olfactory bulb. This phenomenon has been attributed to the fact that the olfactory bulb can detect odors. It would appear that the deeper layers of the olfactory bulb integrate sensory stimuli with centrifugal impulses and, as a result, carry out higher-level data processing to assist in olfactory perception and the control of olfactory-guided behavior.

This would be beneficial for both of these functions. The Odor Code is a system for classifying different aromas. It is mostly unknown how information received through the sense of smell is stored and sent to the brain. As a consequence of this, correlations between electrophysiological responses and stimulus features are weak but statistically significant even though many lines of evidence point in the same direction, suggesting that essential sensory dimensions are unknown (Abdulmajid et al., 2014). There is growing evidence that the mitral cells in the olfactory bulb have odor selectivity. This may be the result of spatially organized projections coming from the olfactory epithelium.

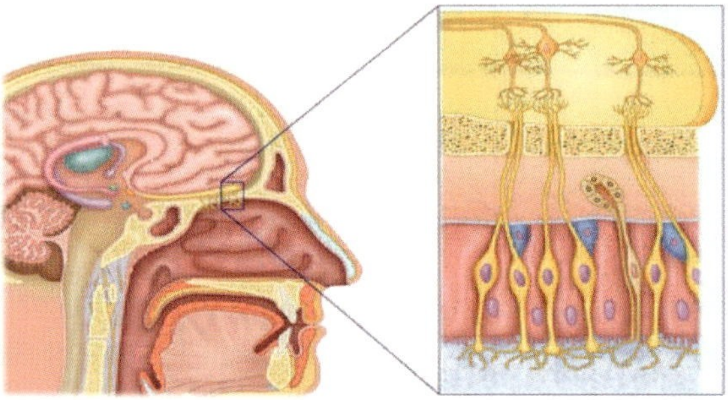

Figure 5.3: Olfaction.

Source: https://t4.ftcdn.net/jpg/02/80/32/57/360_F_280325701_EoDrfNwSJJ-9Hjbp7Mo1I2goedvGRoCMt.jpg.

In addition, the electroencephalogram (EEG) of the olfactory bulb reveals some evidence of olfactory coding. Both learning and habituation have an effect on the putative codes included in EEG and single-unit response data. If these variables are not regulated, differences in the meaning that odorants have for the organism and the impact that arousal levels have on the organism may easily be misinterpreted for odor codes. In a similar vein, the intensity code may be misunderstood as the odor code if the stimulation level for the various perfumes is not properly maintained. These studies incorporated a wide range of variables over which the researchers had little control, which resulted in results that exhibited a high degree of inherent variability. As a result of the lack of a solid basis, theorists are now unable to discriminate between raw sensory information and data that has been processed and interpreted. These signals might be multiplexed by the CNS, which would then create signals that contain interpretative modifiers attached to "words" of sensory information.

Mechanisms for Smell and Taste in the Brain Thompson demonstrates how the parts of the brain that are known to be involved in the control of fundamental behaviors are also the regions that receive olfactory sensory information. Laboratory animals will get electrical stimulation of the medial forebrain bundle as part of the experiment. The hypothalamus plays a significant role in the regulation of appetite, thirst, sexual activity, and sleep (Schiling & Pham, 2012). When the amygdala and the tissues surrounding it are harmed, changes can occur in both sexual and social behavior. Impulses are sent from the olfactory bulb to each of these locations, and the electrical activity of each of these centers is normally almost totally governed by the inputs that are sent from the olfactory bulb.

It is unknown what proportion of the information sent through to these CNS centers consists of sensory input and what proportion has been processed into actionable signals. Signals that are sent to the central nervous system are propagated to a broad range of essential areas, notwithstanding the nature of the signals that are sent. A recent study reveals that the EEG of the human amygdala is connected, in some manner, to the quality of an odor. The effect that diseases of the central nervous system have on behavior is what determines the significance of anatomy and physiology for the field of neurotoxicology. The final and most widespread shared channel is behavior.

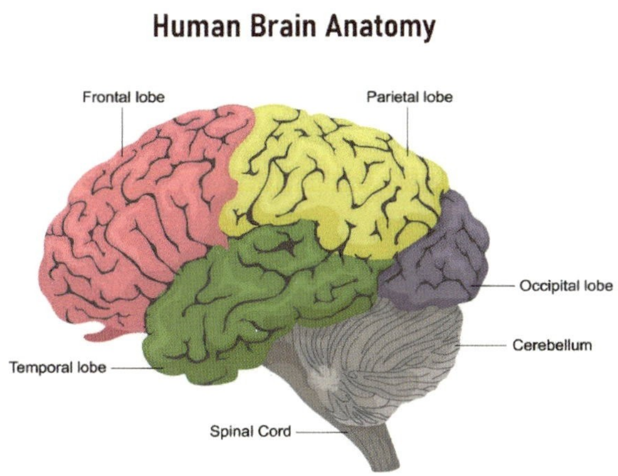

Figure 5.4: The human brain.

Source: https://www.hopkinsmedicine.org/health/conditions-and-diseases/anatomy-of-the-brain.

It is common knowledge that olfactory stimulation can have a significant influence on the sexual behavior of nonhuman species, in addition to social behaviors between and within species, such as aggression, territorial defense, and identification. This is because olfactory stimulation triggers the release of neurotransmitters in the brain. On the other hand, it is generally accepted knowledge that olfactory inputs have only a little influence on shaping human behavior. This assumption is based mostly on the theory about the origins of behavior rather than on empirical facts about the world. Despite this, it is not clear whether or not introspection would provide any important information given the condition of sensory information being conveyed to parts of the limbic system (Shinde et al., 2016). These kinds of subcortical inputs can have an "unconscious" impact. If this is the case, then the impact that odors have on human behavior may be significant, particularly when considering how inconspicuous they are.

It would appear that having an understanding of the effect that odors have on individuals is of the utmost importance in today's society, given all of the efforts that are being made, for example, by modern society to regulate the olfactory environment (e.g., through the use of perfumes and deodorants), as well as the amount of odorous pollution in the environment. It is starting to become clear that the sense of smell plays a significant role in human behavior. Humans may be able to recognize one other by their fragrances, however, the extent to which this occurs is unknown. Pheromone-like compounds are produced by humans, and these chemicals affect both male and female sexual desire. In addition to their role in sexual attraction, olfactory signals may also have a role in other types of social behavior.

There are correlations between a woman's olfactory sensitivity and the periods of her menstruation. In addition, there is evidence to suggest that the scent of jasmine blooms prevents human females from feeding their young. Even though the findings on olfactory impacts on human behavior are just suggestive, such effects appear to be acceptable on anatomical and physiological grounds because connections between the olfactory system and many large CNS areas are created (Sun et al., 2021). To miss the significant yet subtle effect that the olfactory environment has on everyday human behavior would be to ignore the influence that the environment has on people's sense of smell. Regarding this matter, more research is required immediately.

5.3. THEORIES OF OLFACTION

The sense of smell is frequently neglected by modern man in favor of the more prominent senses of hearing and sight. When evaluating this poorly understood sensory system, it is important not to neglect either the purely esthetic value of olfaction in the enhancement of flavor and enjoyment of food or the more important adaptive aspect of olfaction, which allows man to interpret his environment in the absence of visual sense. Both of these aspects contribute to the ability of olfaction to enhance the enjoyment of food. One of the likely reasons for olfaction's relegation to a secondary sense is that it is more difficult to test than the other senses. This is the conclusion that was arrived at by several people, including Kalmus and Hubbard. Olfaction, in the opinion of these experts, is not only a subjective feeling but also a bridge between the objective and the subjective.

Despite these challenges, a significant number of researchers believe that the fundamental benefits of olfactory research continue to be as unknown as they are diverse. In 1964, Snider held the belief that the study of olfactory phenomena could lead to unexpected benefits for humanity. (Serventi et al., 2022). These benefits included improvements in industrial hygiene, advancements in nutrition through the analysis of complex flavors of foods, the elimination of obnoxious odors, the synthesis of any desired odor, and the potential development of an artificial olfactory apparatus or nose as an aid to criminologists and in the anosmic prosthesis, as well as a useful tool in diagnostic Melani As a consequence of this, it would appear that this long-ignored sensory domain requires research that is methodical and in-depth from the perspectives of both pure science and more applicable subfields of psychology.

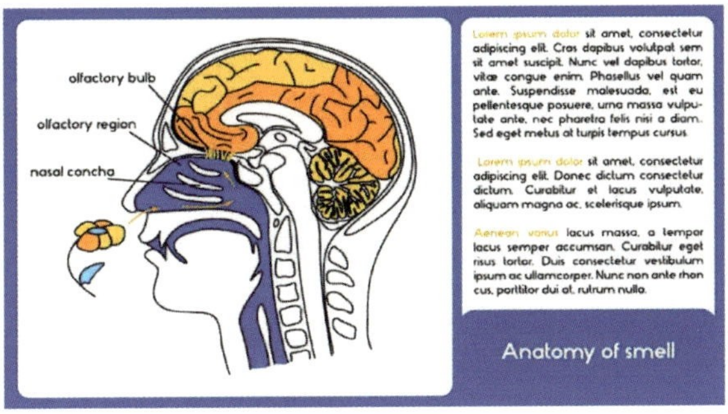

Figure 5.5: Anatomy of smell.

Source: https://thumbs.dreamstime.com/b/anatomy-smell-olfactory-region-to-bulb-olfactorybulb-vector-scheme-104089894.jpg.

5.4. PARAMETERS OF OLFACTORY STIMULATION

It would seem obvious that no investigation of a sensory modality could be considered to be finished in its entirety without first addressing the fundamental stimulus characteristics. In 1953, Jones and Jones conducted research to evaluate several characteristics of odorous chemicals that were thought to be significant for olfactory perception at the time. The following constitutes the majority of the general requirements: It must be slightly volatile at room temperature, slightly soluble in lipoids, at least marginally soluble in water, have a molecular weight that falls within the parameters set forth, among other certain characteristics. According to Jones and Jones, the scientific community does not agree on the validity of these criteria in a unified front. According to Stevens, there is a low probability that a single physical characteristic will affect the form of the appropriate stimulus. According to Stevens, medication must be volatile at room temperature and lip soluble for it to be referred to as having a "smellable" quality (Silva et al., 2020). Odiferous chemicals typically take the form of organic compounds because of their prevalence.

The degree of stimulation that is necessary can be controlled by the drugs' volatility, lipoid solubility, general chemical structure, and spatial arrangements within the many forms of the same molecule. When a person takes a breath in via their nose, the air stream does not travel directly over the olfactory receptors; rather, radiation and convection currents are utilized throughout this process. Despite the relative difficulty of accessing olfactory receptors, odorous chemicals can be perceived even at extremely low concentrations. It is believed that our sense of smell is ten thousand times more sensitive to differences in molecule concentration than our sense of taste. The process using dilution produced threshold values for ethyl mercaptan concentrations that were 4×10^{-5} milligrams per liter of air (Rousta et al., 2021). In general, the absolute threshold for scent varies widely both across and even within individuals, even though all studies appear to agree that the olfactory system is exceptionally sensitive.

5.5. OLFACTION-RELATED PHYSIOLOGICAL MECHANISMS

An organ that detects scent is called the olfactory epithelium, and it is located on the walls of the nasal cavity on each side. Three different kinds of cells make up the olfactory epithelium: pigmented columnar cells, which are responsible for providing structural support; basal cells, which are responsible for producing the substrate upon which the epithelium is created; and olfactory receptor cells. Changes in receptor shape are recognized by Biedler (1961) as being associated with variations in odor sensitivity and, hence, as a possible foundation for distinguishing between odor qualities. The formation of olfactory nerve fibers occurs when the neural projections of these receptor cells come together to form bundles. These fibers arrive at their destination in the olfactory bulb after passing through the bony roof of the nasal cavity.

Because the olfactory bulbs derive from the cerebral hemispheres, the receptors that make up the olfactory neural apparatus have a direct connection to the cerebrum. Because it is connected to several different regions of the brain, the olfactory area of the cortex is one of the brain's contributors to the sensory component of conscious activity. According to Wyburn, the olfactory sense possesses capable neural machinery that makes use of the variables of the impulse code to sort the information that it conveys to the cerebral cortex based on the intensity and variety of the smell attributes of the stimulating chemicals. This information is then processed by the cerebral cortex. It has been established that many types of structures may be found in olfactory receptors. At the surface of the olfactory epithelium, electrical changes created by receptor potentials may be detected when triggered by odorous compounds. These changes can be caused by odorous chemicals (Pu et al., 2020). The magnitude of the electrical responses varies according to the density of the receptors in various parts of the olfactory epithelium, and the recordings show that localized clusters of receptors are particularly receptive to certain odors. This is because the electrical responses are proportional to the density of the receptors. The intensity of the stimulus is what determines the form, duration, and latency time of a reported reaction.

5.6. OLFACTORY QUALITY CLASSIFICATION SYSTEMS

The early attempts to establish theories about the sense of smell depended on classifying different odors. Linnaeus, a well-known taxonomist, was the first person to propose classifying plant odors into the following seven categories: aromatic, fragrant, ambrosial, alliaceous, hircine, and unpleasant. He also proposed that nauseating be its category. In the year 1753, Haller developed a system for classifying fragrances into three distinct groups: pleasant or ambrosial smells, intermediate odors, and offensive smells. In addition to these ideas, there were two categories of odoriferous compounds developed in the 18th century based on the chemical properties of the compounds. One of these classifications was developed by Lorry in 1784, while the other was developed by Foureroy in 1798. Before Zwaardemaker, five less well-known efforts took place in the nineteenth century. In 1895, Zwaardemaker followed Linneaus' categorization approach (Peinado et al., 2014). He added one class from Lorry's list and one class from Haller's list to form a total of nine fundamental odors: ethereal, aromatic, fragrant, ambrosial, alliaceous, empyreumatic, hircine, and nauseating. This brought the total number of fundamental odors to nine.

After conducting a great deal of research, Henning came up with a classification system in 1915 that included six primary categories of odors. These categories were able to be arranged in a prismatic form that came to be known as Henning's Smell Prism. Henning describes the odors as having qualities such as being airy, fragrant, spicy, resinous, charred, and rotting.

In addition, Rimmel constructed a theory that included an unusually high number of different classes. It is important to emphasize that the Rimmel classification system was developed specifically for perfume manufacturers and was not intended as a physiological or component classification of odors. Additionally, Crocker and Henderson put out a theory in which they postulated that there are four distinct varieties of smell nerves, each of which corresponds to one of the following four categories of odor: aromatic, acidic, burnt, and caprylic or goatlike. A single odor can have all four of these fundamental smells, and researchers determined that the relative degrees of stimulation shown by each of these scents were the distinguishing characteristics that contributed to the individuality of the odor.

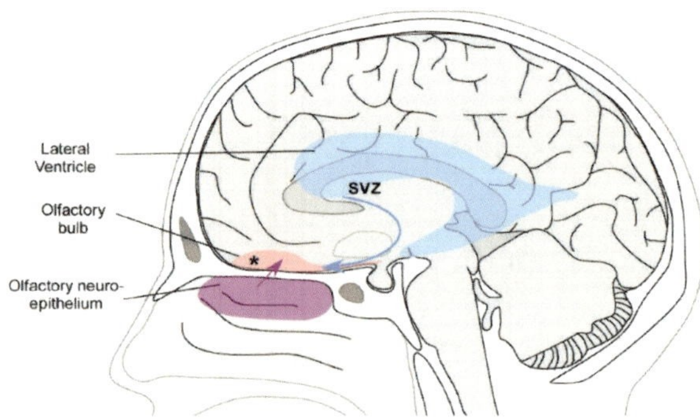

Figure 5.6: Olfactory Bulb High Resolution.

Source: https://www.mdpi.com/1420-3049/18/9/11586/htm.

In addition, Crocker and Henderson (1991) developed the very first numerical coding system to facilitate the process of using their categorization strategy. Recent proposals for classifying scents include the categories of (Duncan et al., 2003). Duncan et al. created a methodology that imitated the process of independently determining thresholds so that he could collect data suggesting the existence of a large but restricted number of relevant odors (Abamecha, 2020). Even though Wendt claimed that these odors are determined by the activation of receptors, he failed to discuss either the nature of how receptors work or Duncan's categories.

Duncan et al. offers a comprehensive taxonomy of fragrances, which may be found on their website. Camphoraceous, peppermint, ethereal, floral, musky, and rotting are some of the categories that aromas might fall into. Jones and Jones, in an outstanding research paper published in 1953, provide a range of reasons for the current olfactory theory to abandon odor classification. This is despite the tremendous amount of work that has been done on this subject previously. The following are some illustrations of this phenomenon: It has never been possible to accurately forecast a substance's scent or even determine whether or not it will have a smell based purely on its chemical makeup. Increasing the concentration of some substances can cause a change in the quality of their odor, while compounds with radically distinct molecular structures can produce odors that are similar to one another, stereoisomers can have smells that are distinct from one another, and compounds that are chemically very similar can have a wide range of aromas. It is important to note that Amoore's current classification has been able to overcome some of these constraints, which is a result of the significant amount of research that has been carried out by Amoore and his colleagues.

5.7. INNOVATIONS IN THE SENSORY RECEPTION OF ODOR

Because of the variety of these thoughts and the limitations of this research, there will be no attempt made to examine them. Those who are interested in this matter can refer to (Bourgeois & Bourgeois, 1970).

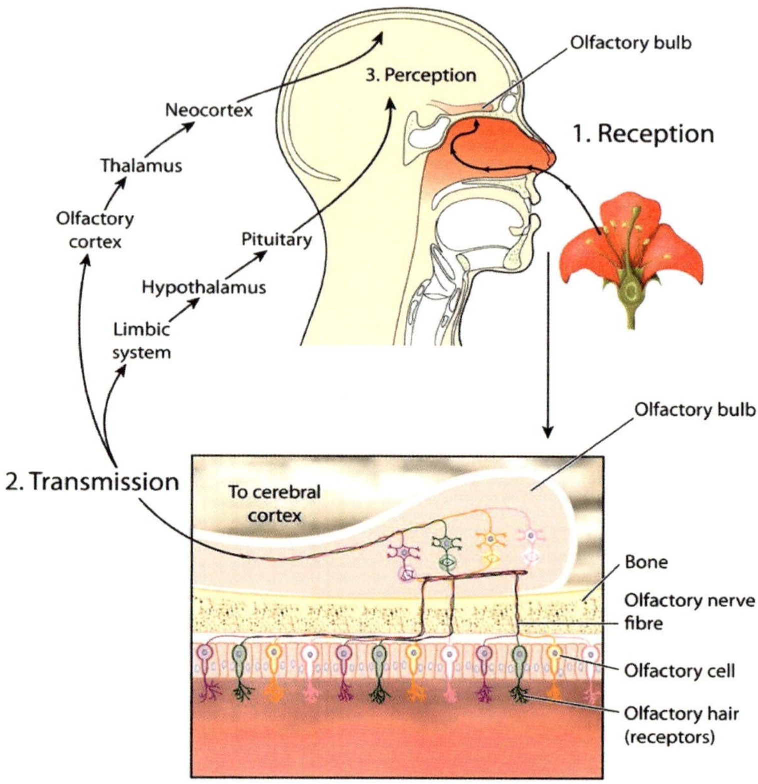

Figure 5.7: Sense smell.

Source: https://thumbs.dreamstime.com/b/sense-smell-transmission-showing-scent-reaching-olfactory-bulb-nerve-signals-passing-to-brain-perception-84909086.jpg.

5.8. INVESTIGATING DIFFERENT CONCEPTIONS OF RADIATION

A learner who is not familiar with olfactory ideas may struggle to understand what is meant by the phrase "radiation." Jones and Jones contend that radiation theories propose that an odor's source emits waves or radiations and that these waves are responsible for the odor. This line of thinking can be shown by the Teutonic hypothesis, for instance. By Teudt's theory, oscillations in atoms or molecules create vibrations in the medium that surrounds them. These vibrations are subsequently amplified via resonance

(Perera et al., 2014). Assuming that the electrical vibrations in olfactory nerves are identical and that there are few differences between them, odor molecules will magnify the nerves whose period of vibration matches their own. This is because odor molecules have their period of vibration. Smells may be classified in several different ways based on their vibration rates. There is no question in my opinion that the validity of these theories has been called into question for a considerable amount of time.

According to the findings of a study carried out by Heyninx, odorous substances that generate bands of ultraviolet absorption are composed of molecules that vibrate at a rate that is proportionate to the wavelength of ultraviolet light that they take in. It could be possible to utilize variations in UV absorption bands to assess shifts in the vibrational frequencies associated with particular odors. According to Moncrieff, hydrogen cyanide has the same absorption bands as water but smells very differently. Jones and Jones point out a variety of additional problems with the UV hypothesis.

5.9. THE SHIFT THEORY, DEVELOPED BY RAMAN

When a substance emits light with wavelengths that are both shorter and longer than the wavelength at which the light was first produced, an optical phenomenon that is known as the Raman Shift takes place. The difference in wavelength that is measured in nanometers between the light that is entering and the light that is being reflected is referred to as the Raman Shift, and it is defined as the difference in wavelength. In 1928, Dyson made his first use of the Raman Shift to get a deeper grasp of olfactory phenomena. This was done so that he could better explain his findings. With the use of the Raman Spectrum, Dyson was able to determine that the exact intramolecular vibrations that could be monitored were the ones that were responsible for odor. The fact that all of these chemicals show a Raman shift in the range of 140 to 350 millimeters implies that they all have the same odor.

Even though it has been shown that the Raman Effect does play a role in olfaction, Wright's theory has been greeted with a large amount of opposition in this field. 1948 was the year that they discovered compounds that had unique odors but shared the same Raman Shifts as other chemicals. Another argument against Dyson's idea is that it has been demonstrated that odorless glycerin exhibits absorption bands on the Raman Spectrum. This finding was made possible by recent research.

5.10. THEORIES REGARDING INFRARED SPECTROSCOPY

During the early stages of the olfactory study, Faraday established a connection between perfumes and the infrared spectrum. According to Glasser, Ogle was the first person in 1950 to put up an infrared hypothesis on how humans experience scent. Based on the fact that auditory and visual receptors respond to waves, Okie suggested that the pigmentation in the nose may as well have been generated by infrared radiation (Abebe et al., 2006). The infrared theory of olfaction is well-known by Beck and Miles. Infrared absorption qualities of odorous compounds cause receptors to lose energy as they reach the radiation zone of the receptors, according to these researchers, since the receptor elements radiate preferentially based on their size and shape. Neuronal impulses rise as a result of the loss of energy.

As a result of receptors radiating selectively and compounds most likely having distinct infrared absorption spectra, certain drugs may activate particular receptors while others will activate different receptors. Several chemicals, according to Morgan and Stellar, show no infrared absorption curves despite their powerful odor. Infrared light is absorbed by carbon dioxide and water vapor, for example. Raman Shift ideas by Young are equally relevant in the infrared as they are in the visible spectrum. As a reaction to these objections, Miles claimed that olfactory material is absorbed as a one-molecule-thick coating on the surface of receptor cells and that cells with an optically active component respond chemically. The initiation of the neural impulse is generated by receptor cells' chemical reactivity. It may be claimed that Beck has rendered his theory inapplicable in the actual world by seeking to adapt it.

In mechanics, theories Banerji's mechanical theory was the sole one found by Jones and Jones in 1953. Odorous particles modify the vibration of olfactory hairs in the nose by altering their molecular weight and velocity, according to this idea. Air moving through the nasal canal is important for odor perception, which gives validity to this notion. This technique has a key flaw: it fails to show a relationship between olfactory quality and molecular weight. Research on macaques' neurological systems discovered evidence of mechanical receptors in the olfactory epithelium.

5.11. THEORIES ON THE PATTERN OF STIMULUS

Several Adrian's research is considered to be among the most exhaustive ever conducted in the electrophysiological field of olfaction. There will be no more investigation of Adrian's findings in this section; instead, Adrian has proposed a new theory about olfactory characteristics. The foundation for this distinction is provided by odor physical features as well as eddy currents in the nasal canal, and it is asserted that distinct odors activate various sections of the olfactory membrane to the greatest extent that is practically possible. As a result of taking into consideration spatial patterning, the durations of arousal and response decay have been changed, and he now believes that this is sufficient to convey quality (Predieri et al., 2021). Other researchers have not been able to replicate Adrian's findings, and the primary criticism leveled against his theory is that it fails to take into consideration the role that olfactory signals play in the process.

5.12. CHEMISTRY THEORY

There is a wide variety of approaches that have been taken in the development of theories regarding olfaction and chemical reactions. To determine the olfactory thresholds of various molecules, including paraffin, alcohols, and chloroparaffins, their solubility properties were investigated. According to him, the olfactory epithelium contains at least two distinct types of receptors, and most likely much more than that; each of these receptors has a unique solubility parameter.

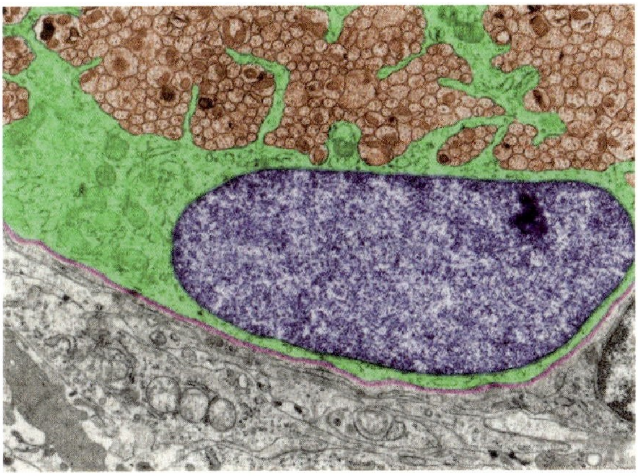

Figure 5.8: Olfactory system.

Source: https://thumbs.dreamstime.com/b/olfactory-nerve-false-color-tem-micrograph-showing-ensheathing-glia-nucleus-blue-cytoplasm-green-basal-lamina-pink-230452646.jpg.

There is the possibility that receptor membranes have a diverse assortment of holes that are randomly dispersed over the surface of the receptor. It is activated when an odorous molecule comes along that is the right size to fit the pore, but pores that are any larger than that stay inactive at this region on the surface of the receptor. As a result of the fact that the concentration required to narcotize some receptor types requires a higher concentration than that required to activate some receptor types, the total responsiveness decreases and adaptation takes place. Large molecules don't produce scents because there aren't enough holes in the membrane for them to fully activate it, therefore they don't pass through the membrane (O'Mahony, 2018). The theory is extremely flawed because it is predicated on the existence of pores that do not exist. The characteristics of receptors are, according to Bourgeois and Bourgeois (1970), the most important aspect of their hypothesis.

A thorough investigation into the yellow lipids found in the olfactory mucosa is carried out. Sadly, most proposals for new mechanical devices only concentrate on a single or a few very minor aspects. A scientist suggests that the sense of smell may be understood in terms of the processes that occur in chemicals. According to Haagen-Smit (1952), the mechanism by which odoriferous chemicals carry out their functions is by disrupting the enzyme-catalyzed reactions that take place in olfactory receptors. Our amazing sensitivity and adaptability can be attributed to the fact that enzymes are affected by very little quantities of a variety of chemicals during their interactions, as Haagan Smit has pointed out. The rapid restoration to the normalcy of the receptor systems would be explained by the fast enzyme inhibitory effects; this would make it possible to detect newly generated odoriferous stimuli. Stoll offers a second solution to explain the odor differences by using a series of synthesized bicyclic farnesyl compounds that are identical to the first.

According to this theory, the intensity and olfactory quality of an odor increase as the molecular weight of the odorant in question increases. He asserts that the structure and size of a molecule, not the functional groups of a molecule, are the most important factors in determining the strength and character of an odor. According to Stall's research, smaller molecules have a greater number of receptors than larger ones, which have a lesser number of receptors (Nishinari & Fang, 2018). Carotenoids were shown to be present in the olfactory epithelium of cows and pigs by Briggs and Duncan (1961), which supported an old theory that carotenoids play similar functions in the senses of sight and smell. According to the findings of Moulton's research, the pigment dispersion in the olfactory mucosa varies between species.

CHAPTER 6

IMPORTANCE OF SENSORY EVALUATION IN FOOD

CONTENTS

6.1. Definition 130

6.2. Applications 139

6.3. Advantages 140

6.4. Disadvantages 141

6.5. Sense Analyzes Concerning Dairy Products 142

6.6. The Aroma and Taste of Food Products 142

6.7. Is Sensory Assessment Useful in the Food Industry? 144

6.8. Dissimilarity Exams 144

6.9. The Descriptive Examination is the Second Type of Examination 146

6.10. Emotional Testing is the Final Stage 148

6.11. Morphological Characteristics 149

6.12. Blinds With Multiple Colors 152

6.13. Summary of Sensory Testing 153

6.1. DEFINITION

Perceptions of food and materials are considered visually, auditory, tactilely, and olfactory. Food, participants, the testing environment, and the evaluation methods themselves all have an impact on sensory evaluation (Nwakalor & Obi, 2014).

6.1.1. Objectives

An effective sensory evaluation program must include goals and objectives, a well-defined program strategy, professional personnel, and appropriate testing facilities. The sensory evaluation seeks to accomplish the following objectives:

- To understand how people perceive dairy products based on their taste and smell.
- To provide timely and relevant recommendations and information on the sensory evaluation of dairy products (Muhimbula et al., 2011).
- The development of methods and procedures for the correlation of sensory and analytical data benefits research, quality assurance, and quality control.
- To demonstrate how dairy products can be evaluated based on flavor and texture.

Figure 6.1: Sensory evaluation.

Source: https://media.istockphoto.com/photos/professional-asian-chinese-barista-grader-performing-coffee-cupping-picture-id1338803468?k=20&m=1338803468&s=612x612&w=0&h=oZNy0FcNHmkTVJ_SWw9af6m7Re3W-fR25AEOBwBa_bYs=.

Sensory analysis can be used to identify the attributes that drive consumer preferences in addition to quality control, shelf-life determination, product readiness for market introduction, and flavor profiling. This tool makes it easier to make critical decisions about raw materials, ingredients, additives, storage conditions, expiration or "best before" dates, and product optimization (Munasinghe et al., 2013). When the sensory analysis is used, it can provide valuable information and insights that can be used in the evaluation of food products.

- Ensure that the customer's expectations are met or exceeded.
- Product questions that have a direct impact on revenue and market success should be addressed.
- Analyze consumer feedback to make well-informed purchasing decisions for both current and future products.
- Look into any problems that may arise.
- Comparing different brands can provide a competitive advantage.

Food sensory analysis includes the description, measurement, and interpretation of product characteristics perceptible to human senses. This is an example of a measurement technique that, like other instruments, employs a human subject in the lab. The sensory analysis aids in distinguishing between two types of tests. The goal of objective sensory analysis, also known as analytical analysis, is to generate data that can be compared to data obtained using chemical or physical measuring equipment. Popularity or preference for a product, on the other hand, is a subjective impression that is emphasized in hedonic evaluation. We have two types of sensory testers: trained and untrained because trained individuals conduct objective tests and untrained consumers conduct subjective tests (Mouta et al., 2016). To avoid the limitations of sensory evaluation, the food industry and academia employ a variety of analytical methods for detecting hydrolysis and oxidation molecules in foods. Even though these techniques are not directly related to food sensory quality, they provide objective measurements and potential insight into the cause of flavor quality declines.

Only a few methods can account for every variation in any food system. Sample preparation is required for gas chromatography measurements of volatile compounds in food samples and solvent extraction measurements of peroxide in snacks. Analytical and sample preparation methods must be customized for each food matrix. Sensory evaluation has applications in many fields, including research, marketing, and quality control. Sensory science research into product differences, consumer preferences, and attribute intensities benefits both new product development and product improvement. This information is used to identify consumer segments, validate advertising claims, and keep tabs on the competition. Sensory quality control can be used to ensure that finished goods are consistently produced. Sensory evaluation can be used to assess the effects of processing or ingredient changes, plant-to-plant variation, pilot-plant scaling, cost savings, and packaging modifications. Certain sensory properties may influence analytical or instrumental analyzes (Lomarat et al., 2019). Finally, sensory evaluation can help you make sound business decisions based on the most recent data. To minimize the impact of external factors such as weather, guidelines for the preparation and serving of samples under controlled conditions are required for the science of sensory evaluation. Through the quantitative science of sensory evaluation, human perception and product attributes are mathematically linked (Lawless and Heymann, 1998). Human observers collect data with many variables, and human panels are by definition heterogeneous instruments.

Statistical analysis is used to determine whether the correlation between product characteristics and sensory responses is real or the result of uncontrolled response variation. When using descriptive sensory evaluation, an evaluation panel must be able to identify a wide range of sensory characteristics and agree on how to label them. Assessors of descriptive sensory evaluation must receive extensive training before they can achieve the panel's required level of objectivity and reproducibility. When using a panel for an extended period, it must be maintained and calibrated regularly. In sensory evaluation, which is a quantitative science, human perception and product characteristics are inextricably linked. Our perception of the intensity of certain sensory characteristics changes as we age. The goal of time-intensity sensory evaluation is to track how an attribute's perception changes over time. Tastants such as sweet and bitter stimuli, as well as flavor release phenomena in food, have been extensively studied using time-related methods.

Figure 6.2: Food evaluation.

Source: https://www.medallionlabs.com/wp-content/uploads/descriptive-sensory-evaluations.jpg.

When it comes to sensory evaluation, "evocation, measurement, and analysis" are all utilized. According to numerous criteria, including price, desire, prior experience, and what is perceived to be 'healthy' people make their food choices depending on their preferences. But more crucially, they employ information obtained from their senses: sight, touch, taste, and odor. A food's esthetics and quality may be increased with the use of certain sensory components, which may also be customized to match the tastes and dietary requirements of diverse populations. The sensory assessment of food items can supply manufacturers with data for product development, marketing, and other aims.

The objective evaluation of food's flavor, smell, and texture by the use of human senses is known as food sensory analysis. Product quality is examined, concerns are rectified, and new things are produced utilizing this process. A wide range of sensory analysis methodologies has been mastered by our staff (Lee, 2013). The relevance of sensory assessment in food testing is progressively becoming acknowledged. It is becoming increasingly essential for ensuring the quality and commercial success of food products. Many stages of product development and manufacturing can benefit from it.

Ensure that testing is undertaken regularly and that data is collected in a speedy, efficient, and consistent manner to enhance the effectiveness of this sort of study. Data must be given in its totality, and its quality must be maintained throughout time. The analysis may be done by a third party or internally with the help of an expert. To make the most of your sensory analysis, you must engage with a trustworthy partner.

Not only is sensory assessment crucial for quality control of manufactured things, but also product creation. Sensory evaluations are done by professionals with varying degrees of knowledge and experience. Untrained folks, such as supermarket shoppers, are employed as assessors when given food samples and requested to provide an "overall" opinion or to define what they appreciate or despise about the cuisine they are eating. The second sort of assessor has gained particular training in food sensory evaluation. Lastly, there are specialized assessors, who are highly educated individuals with a profound understanding of product texture, consistency, visual appearance, flavor, and taste. Statistical mathematics, as well as an examination of correlations, degrees of dependability, and levels of relevance, are important to perform a sensory evaluation quickly.

Figure 6.3: Sensory analysis.

Source https://www.istockphoto.com/photos/sensory-evaluation.

Sensory assessment is a vital way of determining the taste quality of food since it investigates what customers genuinely feel; yet, it is a subjective procedure. For instance, one buyer may find a sample disgusting while another finds it appealing. These variances in sensory perception are ubiquitous and can be explained by nationality, culture, and individual diversity. To increase the impartiality of sensory evaluation, a sufficient number of panel members must be trained (Lee et al., 2021). Techniques and parameters for training must be methodically constructed depending on the intrinsic properties of the examined food matrix, as well as any potential variation between samples arising from food formulation, processing, packing, and storage conditions, among other variables. Consequently, sensory testing may be time-consuming and costly.

To escape the constraints of sensory evaluation, the food company and academics utilize a variety of analytical procedures to discover hydrolysis and oxidation components in meals to verify their stability and quality. Although these techniques are not directly related to the sensory quality of food, they give objective measures and likely insight into the reason for taste quality decline. However, no one analytical approach can account for all changes in any food system. For the majority of analytical procedures, sample preparation is needed, including appropriate heating for gas chromatography measurements of volatile chemicals in food samples and solvent extraction for peroxide tests in snacks. Consequently, an appropriate analytical method and sample preparation approach must be selected for a specific food matrix. In earlier chapters of this book, the details of a range of analytical methodologies for measuring hydrolysis and oxidation components in diverse food matrices were explored.

A scientific domain that elicits, evaluates, analyzes, and interprets reactions to the properties of food and other substances as experienced by the senses of sight, smell, touch, and taste. Sensory science is incredibly important to the consumer goods industry. In addition to studying foods and drinks, sensory science is also utilized to examine personal hygiene items, home goods, and ambient scents. Even though analytical techniques may be utilized to quantify several food and beverage properties, sensory science, which employs human assessors, gives a unique insight into the sensory experience of specific meals and drinks. Research, marketing, and quality control can all benefit from a sensory assessment. To learn about product variances, consumer preferences, and attribute intensities, sensory science is employed in the production of new items and the refinement of present things (Lotfy et al., 2021). Marketing departments employ sensory assessment data to develop customer categories, support advertising claims, and track competition. A sensory quality control department may examine product maintenance to assure consistent creation of final items. Sensory assessment can be used to analyze the impacts of processing or ingredient adjustments, plant-to-plant variance, pilot-plant scale-up, cost savings, and packaging enhancements. Specific sensory characteristics may be connected with analytical or instrumental research in specific circumstances.

Figure 6.4: Food sensory evaluation.

Source: https://media.istockphoto.com/photos/barista-preparing-to-test-and-inspecting-the-quality-of-coffee-picture-id1284116251?k=20&m=1284116251&s=612x612&w=0&h=8C45ItEKLFswYpmsy8XJ2geyH2KSmwOineFyUfzt-few=.

Due to its effect on product quality as a whole, the sensory quality of food items has been recognized as an important component since the beginning of the food industrialization process. In terms of sensory qualities, quality is measured by the presence of appropriate levels of sensory properties such as sight, smell, taste, and texture. Sensory analysis is utilized to characterize and quantify a product's sensory qualities. Sensory analysis is the scientific examination and description of the features of a product as perceived by the five senses: sight, hearing, smell, and taste. Understanding sensory data helps you to advise food-product designers on the traits to highlight when choosing things. This decision is governed by processing components and cost factors (López-López et al., 2019). In addition to "tasting" the meal, it may include characterizing its color, texture, flavor, aftertaste, scent, tactile reaction, and even sonic response. Sometimes, sensory analysis and sensory assessment are used interchangeably.

The study of nature relies heavily on sensory analysis. Food sensory aspects must be measured with caution. Sensory data, when properly analyzed, can provide a wealth of information about one's surroundings. Incorrect measurements cause more harm than good by concealing the truth. During sensory analysis, sophisticated controls such as neutrality in sample presentation, reduction of response bias, and the use of methods that require panelists to demonstrate their abilities rather than relying on self-reports must be designed and implemented. When these rules are broken, the resulting sensory data becomes less useful. In contrast, establishing and enforcing appropriate controls will yield meaningful data on the nutritional value of foods that cannot be measured by any instrument.

Figure 6.5: Sensory analysis.

Source: https://media.istockphoto.com/photos/human-takes-a-heap-of-roasted-coffee-beans-by-both-hands-from-a-bag-picture-id1303762845?k=20&m=1303762845&s=612x612&w=0&h=Zo_YOz5WgH2Y59RpDj27s6UVTdnEJX-uthnypx_U4Dpk=.

The sensory characteristics of a food product can be used to aid in product development, marketing, and other business operations. This evaluation considers all aspects of a product's sensory experience, including aroma, flavor, and texture. The foods we eat are influenced by sensory factors such as smell, sound, sight, and texture. To provide the best possible eating experience, food should have a pleasing aroma, flavor, texture, and mouthfeel (Kweku Amagloh et al., 2013). Food sensory association analysis knowledge includes the aforementioned outline; it aids in the measurement and interpretation of product characteristics that can be observed by human sensory organs.

In sensory science, which is a subset of food science, there are numerous methods for accurately assessing human reactions to food.

In terms of intelligence, a machine cannot match the quality of a human brain. As a result, in sensory science, human subjects are used as test subjects. Although electronic tongues and noses have made significant advances in measuring taste and smell, they can only focus on one sensory component at a time, whereas the human brain can absorb and respond to multiple sensory properties at the same time.

Consumers base their food choices on several factors, including the food's health and safety, price, availability, and even sensory qualities such as color, texture, flavor, and aroma. Because customers' needs and desires are constantly changing, sensory science is essential in predicting how they will react to various products. Food manufacturers research these ingredients to ensure that their products are of the highest quality and profitable. The importance of consumer perceptions of food acceptability and pleasure emphasizes the importance of sensory science in the food industry. Sensory food science may be able to assist food companies in meeting their strategic research and financial goals.

Sensory science, in addition to shelf life research, can be used to identify factors that influence consumer preferences and to improve product development and optimization. It is used to assess the overall quality of food and drink, in addition to meeting legal and safety requirements. Sensory and consumer data are required to establish quality standards for products that are important to customers. With the help of this data and information, informed decisions about the storage and packaging of raw materials, components, and additives can be made (Koli et al., 2022). There are numerous non-food applications for sensory data generated by sensory testing. Did you know that humans can be used to improve animal feed quality? Participants in the food industry rely on food sensory evaluation insights because they can be used to:

- A business must fulfill its clients' requirements and expectations.
- You may generate more money by enhancing or developing new products.

- Troubleshoot and correct any issues with the food or the manufacturing methods.
- Gain an advantage over your competitors in the same market.

To achieve a competitive advantage in the market, food companies are increasingly incorporating sensory analysis into various stages of the production process. The purpose of sensory assessment is to offer reliable and trustworthy data to research, manufacturing, and marketing departments so that management may make well-informed decisions on the perceived sensory qualities of the product. Matching sensory features with scientific, physical, or chemical research can save money. Aspects like the following also have a key impact on sensory judgment:

A man's preferences and dislikes concerning dairy products are well-developed. Sensory assessment will become increasingly crucial as the ordinary customer has greater awareness of nutrition and quality. Evaluating a food's edible characteristics is made easier through sensory evaluation (Kukovics & Németh, 2014). The integration of objective and sensory data is vital to acquiring the most accurate information. Smells and tastes may be recognized even when there is no conversation between the participants. The Six Senses give a complete perspective of brilliance.

The sensory assessment gives unique and incredibly crucial information in the market. If or whether the program is effective will be judged by the amount of skill and ability to influence decision-making demonstrated by the program's participants, and sensory professionals.

6.2. APPLICATIONS

Sensory analysis can be used to answer the following questions.

- What does the product taste like?
- What are its sensory characteristics?

Changes in a product's manufacturing, packaging, or storage can all have an impact on its sensory attributes. Sensory assessment may be used in the dairy business to achieve the following goals:

- A competitor's product might be compared to the company's own.
- Objects can be improved by modifying or replacing their constituent elements.
- Ascertain that all standards are satisfied.

- To guarantee quality control, compare samples regularly to requirements.

Figure 6.6: Sensory analysis.

Source: https://media.istockphoto.com/photos/wine-critics-at-work-picture-id479996984?k=20&m=479996984&s=612x612&w=0&h=V5aJ3MSNWj9a3VBALpT1ze-f7kC1vd467mkxHH21qv4=.

It is critical to be able to identify between batches or manufacturing runs. The characteristics of a new product should be profiled. In your description, consider the product's unique traits. Show the marketing staff new items. Customers must be informed about new or reformulated goods.

6.3. ADVANTAGES

- Assists in gaining a better knowledge of regular consumers' interactions with food goods.
- The following consumption, the product's overall impression, i.e., its eating quality, is rated. Why does non-sensory panel testing take so long compared to sensory panel testing? Because they employ so many senses, they are more versatile as instruments. Its high sensitivity implies that it can detect even minor changes in product quality.

Figure 6.7: Sensory analysis of food.

Source: https://media.istockphoto.com/photos/wine-tasting-event-picture-id491411072?k=20&m=491411072&s=612x612&w=0&h=odsOEYr3FeZWGb_fuFzaOpTfpHDaxZtKMY4alG35gKU=.

6.4. DISADVANTAGES

- Testing and evaluating descriptive data might exhaust sensory panelists.
- Assessors' judgments may be biased as a result of boredom or diversions.
- A team of assessors may be necessary for statistical and descriptive data analysis, making this a costly effort.
- This procedure can be costly and time-consuming.
- A new assessment will require substantial training to have the necessary competence for the post, which may make it difficult to replace existing assessors fast.
- Sensory panels can be more expensive than non-sensory solutions.
- The panelists' perceptions may be inaccurately measured.
- Interpretation of findings can be difficult and contentious.

6.5. SENSE ANALYZES CONCERNING DAIRY PRODUCTS

The dairy business was one of the early adopters as a consequence of sensory analysis. The sensory assessment of dairy products comprises a detailed investigation and interpretation of the product's primary sensory qualities. In the early 1900s, methods for evaluating dairy products were created to stimulate public interest and teach people about dairy science.

Chemical procedures, microbiological tests, and sensory inspection can all be used to track them down. A sensorimotor assessment is one of the simplest and quickest approaches. A food technologist who specializes in dairy flavor analysis has an edge over someone who just understands how to test goods chemically or microbiologically. It is vital to precisely identify the kind and cause(s) of sensory abnormalities to employ remedial procedures throughout the manufacturing, processing, and distribution phases (Kohyama, 2020). As the first stage in a quality assurance procedure, dairy processors must conduct a taste test on all dairy products. Sensory assessments may result in the rejection of some varieties of milk, cream, and other dairy products. One of the most widely held industry beliefs is that dairy products are only of excellent quality if they are created from high-quality components.

When judging and grading dairy products, quality scores are frequently offered by one or two trained "experts." The existence or absence of predetermined flaws influences how well an item's look, flavor, and texture are assessed. Traditional techniques of product quality evaluation may be beneficial in a fast-paced industrial context, but they are useless when it comes to new product creation and the manufacture of new goods to fulfill market demand. The Bureau of Indian Standards established criteria for rating and assessing dairy products (BIS). Sensory assessments of dairy products are increasingly being used by researchers in their efforts to develop novel products and methods.

6.6. THE AROMA AND TASTE OF FOOD PRODUCTS

Quality-conscious customers are increasingly influencing food design, manufacturing, and marketing. Consumer views of food quality are rapidly changing, emphasizing the necessity for food producers to be inventive to compete. Customers are the most likely to acquire a product since they are the ultimate users; hence, customer acceptability is a factor in a product's success. Customers' opinions on a new product, a formulation change, or a process are therefore highly valued by industry specialists. If people dislike a product after it has been made or evaluated, it is labeled a failure.

Foods can have intrinsic as well as external sensory qualities. Customers may sense physical qualities and utilize them as quality indicators; therefore it is feasible to determine a product's essential traits even before a client eats or uses it. The look, color, form, size, and structure of milk products are all critical factors to consider. An object's physical attributes are intrinsically tied to its core characteristics. Price, brand, production information, nutrition details, package design, place of origin, retailer, and ease of access are examples of extrinsic quality indicators (Kang et al., 2009).

How might engaging people's senses with new items impact their eating habits? All tactics for eliciting, measuring, analyzing, and interpreting human reactions to food attributes as perceived by the five senses are covered in the field of sensory assessment. Taste and scent have grabbed the interest of scientists when it comes to ingestive behavior.

Figure 6.8: Food sensory test.

Source: https://media.istockphoto.com/photos/olfaction-test-female-scientist-examining-vanilla-smell-picture-id1320929449?k=20&m=1320929449&s=612x612&w=0&h=Wl2_57_iYo8SO4vsvcVSi3yOFkNot1ev5dvDZaiScEY=.

6.7. IS SENSORY ASSESSMENT USEFUL IN THE FOOD INDUSTRY?

The following issues necessitate sensory examination:

- the development of new items
- effectiveness in lowering costs
- raising the general population's level of living.
- acceptability of the product
- quality control and management

A sensory evaluation focuses on objective assessments of a product's sensory qualities, subjective responses to physical items, and interpretation of customer responses via product comprehension. Sensory analysis may help with quality control, shelf life prediction, product launch preparation, product success evaluation, taste profiling, and identifying the elements that influence customer preferences (Jain et al., 2014). It might be a useful resource in terms of raw materials, components, additives, storage and packaging, expiry or "best before" dates, and product optimization.

6.8. DISSIMILARITY EXAMS

Food products are inspected during variation testing to see if they differ in any way. Scent, taste, and texture are examples of these characteristics.

- The Triangle Test determines whether two objects have discernible sensory differences.
- The Duo-Trio Test determines whether two objects have discernible sensory distinctions. It is especially important when product changes have occurred as a result of production changes. Each of the three samples is examined by the panelists to determine which is the most distinct. When there is significant flavor carryover between samples and panelists are confused by having to evaluate three models rather than two, a triangle test may not be the best method.

Figure 6.9: Sensory evaluation.

Source: https://media.istockphoto.com/photos/month-old-toddler-with-picture-id1144052881?k=20&m=1144052881&s=612x612&w=0&h=gUg-M02yW-477PYzaxn9QPLL-O32eQ_2ahSUkN-C2IYM=.

- In addition, the Duo-Trio Test can detect product variations caused by changes in ingredient suppliers, storage conditions, packaging, and other variables. The sensory Eater detects a sample that is identical to a reference sample. It is a simple concept to grasp.
- Analysis of Comparability Tests are performed to determine which of two samples has a greater quantity of a property or which of two occurrences is preferred. It is referred to as an acceptance test in the latter case. It is one of the most commonly used attribute difference tests, and panelists can quickly grasp it (Igarashi et al., 2010).

6.9. THE DESCRIPTIVE EXAMINATION IS THE SECOND TYPE OF EXAMINATION

A descriptive food sensory analysis provides a comprehensive profile of a food product's sensory qualities, as well as a qualitative assessment of the intensity of each characteristic.

- Flavor Profile Evaluation. The five essential components of the Flavor Profile method (the overall impression of the analyzable and non-analyzable flavor components) are character characteristics, attribute intensity, the chronology of attribute presentation, aftertaste, and amplitude (the overall impression of the analyzable and non-analyzable flavor components). The flavor profile was first evaluated on a five-point scale as follows: absence, threshold, mild, moderate, and strong. The scale has grown over time to include more points to accommodate a greater intensity disparity (initially seven, now 14). (Originally seven, now fourteen).

6.9.1. The Texturometer

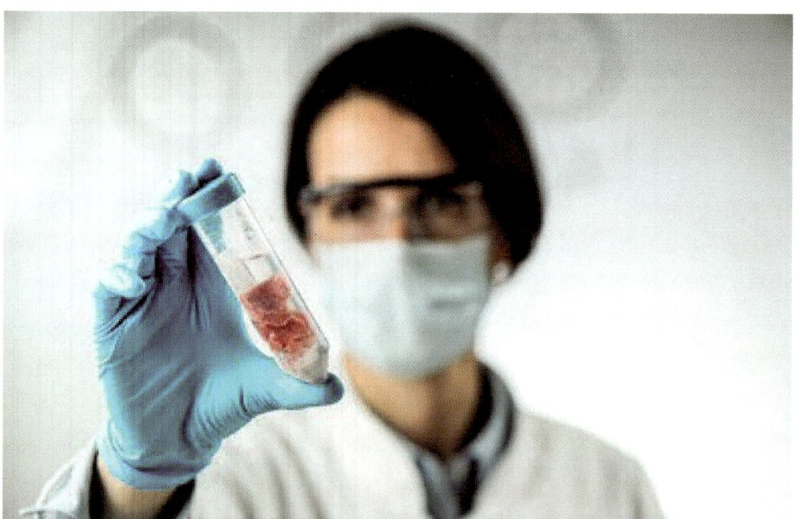

Figure 6.10: Food sensory image.

Source: https://image.shutterstock.com/image-photo/food-quality-control-red-meat-260nw-1943698204.jpg.

- A New Instrument for Objective Texture Measurement included five key food texture characteristics (hardness, cohesion, adhesiveness, viscosity, and elasticity), as well as three additional variables (brittleness, chewiness, and gumminess) (brittleness, chewiness, and gumminess).
- The Spectrum Descriptive Analysis combines the Flavor and Texture Profile Methodologies' rigorous training and organization with descriptive data, as well as a larger panel (up to 15 individuals), a more precise scale (typically 150 points, depending on the product), and statistical tools.
- Qualitative Tests: Ten to twelve people are asked to quantify product features on a six-inch line scale with a half-inch intensity indication.
- Customized profiling: The Right to Choose Profiling is distinct from previous efforts in two ways. Each panel begins with "untrained" clients. Even if they are given evaluation instructions, they are not chosen based on their aptitude or ability to detect minor differences in product quality (Huang et al., 2022). Second, rather than providing an empirical evaluation, participants assess each aspect by responding with "liking" or "acceptance" (among other qualifiers) (among other qualifiers).

Figure 6.11: Taste testing.

Source: https://media.istockphoto.com/photos/professional-barista-testing-the-smell-of-a-new-coffee-picture-id1300058868?k=20&m=1300058868&s=612x612&w=0&h=e7e6zb0FnBNKVv1CL8Po4u4KC3U0PeF5E_GYc1DTmV8=.

6.10. EMOTIONAL TESTING IS THE FINAL STAGE

The testing procedure is clear and straightforward, so the consumer panel will understand what to do.

- The Paired Preference Test assesses the preferences of two people. A preference test can be performed if a significant difference between the two wines has been established. This will help you decide whether to use a wine mix or yeast fermentation.
- Preference Ranking Test: A preference ranking test can be performed if more than two samples are evaluated. In most cases, a consumer can evaluate three to five items in a reasonable amount of time. The client is asked to rank the samples in order of preference, with "1" being the most desired, in this sensory assessment method (Hong et al., 2008).

The Hedonic Test is used to assess a person's happiness level. The hedonic scale can be used to assess the acceptability of a wide range of objects. This scale has an unusually large number of categories, ranging from "strongly dislike" to "strongly enjoy." There is a happy medium (neither like nor dislike) (neither like nor dislike). Customers rate products based on their experiences.

Our taste buds, which number thousands, play the most important role in food selection. Sensory elements such as smell, sound, sight, and touch, in addition to taste, influence what we eat. Food must have a pleasant mouthfeel, texture, appearance, and aroma in addition to flavor.

"The aroma component of flavor is essential," says Jean-Xavier Guinard, program director of the applied sensory and consumer science degree program at the University of California-Davis Extension, which is now accepting applications. Foods must be the correct color, size, shape, consistency, and opacity, as well as smell fresh or ripe (Happiness et al., 2011). Examining foods for flakiness, moistness, dryness, oiliness, and other qualities can reveal a lot about their texture and ripeness, for example.

Figure 6.12: Sensory evaluation.

Source: https://media.istockphoto.com/photos/close-up-professional-asian-chinese-barista-connoisseur-performing-picture-id1343280313?k=20&m=1343280313&s=612x612&w=0&h=p4niJviLCLRTzV3l8aB6Fj8uGLUjPlEx7L-YjgdGIZM=.

Foods must maintain a precise level of crunch, bubble, sizzle, pop, snap, and crackle while retaining shelf life and nutritional composition, so the sound is critical. Crispy and crunchy meals should always have the same crunch and crispness.

6.11. MORPHOLOGICAL CHARACTERISTICS

"Texture is just as important as a flavor because it decides directly whether a buyer enjoys something," says Layo Jegede, senior manager-global sensory at Ingredion's R&D center in Bridgewater, New Jersey. "It is critical to prioritize texture over flavor when developing new products because the texture has a direct influence on flavor release and customer impression."

He is also aware of his other senses. "Before consumption, consumers may "see" flakiness in baked goods. Consumers will not taste a product if it has an unpleasant aroma."

According to Whaley, the number of North American products with a texture claim on the front of the box increased by 74% between 2011 and 2015. Creating different food textures also necessitates a thorough understanding of ingredient structure/function relationships, ingredient interactions, and processing conditions, as well as how they influence sensory characteristics (Gunaratne et al., 2019). "Knowing the properties of starch and the concentration at which it is used allows food producers to control and optimize textural features," she says.

After determining the proper texture, quantitative textural properties can be established and evaluated. "Simple texture modifications may transform a single product into a platform that appeals to a wide range of customers," says Schleicher of TIC Gums. "Manufacturers can create hard, soft, and liquid variants of the same candy by changing the ingredients in their formulas. Gum acacia and other stabilizers change the application's desired texture, allowing formulators to improve and expand their product line."

The selection of evaluation processes, according to Jegede, is dependent on the objectives for a new food product, the given sample amount, and the timeframe. "We use a wide range of methodologies and methods, such as descriptive analysis, temporal perception, oral processing, discrimination, and consumer tools. The emphasis is on sensory quality factors that influence customer acceptance and purchase decisions. Flavor and texture profiling, as well as consumer research, are carried out in the same way that the client or foodservice operator will prepare the product."

Customers update products to differentiate them from the competition or to retain key sensory characteristics of an existing product while making claims such as low-fat, low-sugar, and gluten-free, according to Marcelo Nichi Sr., marketing manager for texture at Ingredion "They are also interested in experimenting with new textures (Fiorentini et al., 2020). There has been a surge in interest in clean-label starches in recent years, and many clients are looking for more cost-effective alternatives to food-modified starches. Understanding these characteristics is critical because they can affect the commercial success or failure of a product."

Figure 6.13: Taste testing.

Source: https://media.istockphoto.com/photos/cup-taster-girl-tasting-degustation-coffee-quality-test-coffee-picture-id1307406406?k=20&m=1307406406&s=612x612&w=0&h=_3AOSxVH6dSKj7Uh33Rvn3sdl0kfSA8K2BYCrl-0hLNE=.

Hydrocolloids and carrageenan, according to Scott Rangus, president and CEO of Ingredients Solutions in Waldo, Maine, provide a variety of textural and sensory properties that can be measured in terms of gel strength and flexibility. "A texture analyzer evaluates the gel break force and elasticity of gelled substances to determine this," he explains. "These procedures ensure that the final consumer products, such as pudding, gummy candies, and dessert gels, have consistent qualities. Dairy drinks, for example, can have their viscosity measured."

Rangus recommends tasting panels in addition to quantitative evaluations if statistics alone are insufficient to distinguish one formulation from another. "Because flavor and mouthfeel characteristics may differ between two items," he observes, "customers are frequently required to make the final determination." "Carrageenan gel, for example, can change the color of deli-sliced turkey lunchmeat (Feng et al., 2019). The resulting flesh may be blackened if a darker gel is used. The same can be said for taste and smell."

Food manufacturers, he says, also want to improve texture and "richness," or the perception of high-fat content in low-calorie meals. Gums can compensate for the loss of body and mouthfeel when liquids are reduced in fat or carbs.

Figure 6.14: Sensory analysis.

Source: https://media.istockphoto.com/photos/selective-focus-of-coffee-shop-workers-checking-coffee-quality-during-picture-id1303762877?k=20&m=1303762877&s=612x612&w=0&h=4KCZpsP2rSVXBzEGDArZn5-Z5h5BRhi-joZc213qVzXM=.

Rangus claims that when a foundation formula is created, it is possible to identify areas of deficiency and seek components to correct such deficiencies. "The process is repeated until the desired texture and sensory properties are obtained. Then, we work with the client to fine-tune the final product."

6.12. BLINDS WITH MULTIPLE COLORS

Color can also convey a lot of information. Indicate whether a dish has been cooked, is burnt, stale, past its prime, or is not yet ready to eat to increase or decrease acceptability.

"Consumers want components they can quickly recognize and feel comfortable with in terms of hue," says Tom Schufreider, COO of Sethness Products Co. (www.sethness.com), Skokie, Ill. "Finish product colors and tints should resemble natural colors or colors produced by traditional techniques such as grilling, roasting, and toasting," he writes. "Some buyers prefer and value items that are more subdued and less vibrantly colored than those that were previously dyed with brighter FD&C dyes (Elsamani et al., 2014). Customers must perceive a manufacturer's palette colors to be more genuine in hue and value."

Mars Inc. is attempting to reduce the number of artificial colors, salt, and sugar in its products, but is finding it difficult in some cases to switch to natural colors. Skittles are famous for their bright reds, greens, yellows, and purples, which are prominently displayed on their packaging and in their "Taste the Rainbow" marketing campaign.

Natural colors that work well with the Skittles recipe may be difficult to find, especially given the rough surface of the candy. Working with suppliers to develop new chemicals and formulations would be advantageous, as the company currently uses a variety of colors derived from biological sources in its global product line. Various formulations and products may exist in various markets based on customer preferences, ingredient availability, and local legislation. However, according to the company, everything is safe.

"Product developers must thoroughly examine the colors they want to use to ensure that their performance and long-term integrity are not jeopardized by interactions with other components," Schufreider explains. "A finished product's pH, alcohol concentration, salt content, protein content, and ionic charge, for example, may affect the apparent color, performance, and durability of caramel colors (Elina et al., 2016). Knowing what to expect in terms of interactions may help prevent color instability or predict desirable hue changes."

Sensory testing can give you valuable information about how your customers perceive your product through their five senses, allowing you to enter their minds. By combining the five basic senses of taste, touch, sight, and sound, we can cultivate subconscious brand loyalty.

Similarly, what does sensory assessment entail? Sensory analysis (or sensory assessment) is a scientific field that combines experimental design and statistical analysis to evaluate consumer items using human senses (sight, smell, taste, touch, and hearing) (sight, smell, taste, touch, and hearing).

6.13. SUMMARY OF SENSORY TESTING

Sensory evaluation is currently underutilized in the food and beverage industry, even though it may benefit a vineyard that employs the techniques efficiently. Sensory assessment can serve as a basis for judgments if controls are properly established, panelists are carefully selected, practical training is provided, sufficient sensory procedures are used, and trustworthy statistical interpretation is undertaken. Food Research Lab's expert sensory panelists assist clients in studying and selecting the best food. Food Research Lab is a global contract research and development food, beverage, and nutraceutical lab that provides solutions to the F, B&N industries. We are one of the world's most well-known food sensory testing companies.

Increased competition and new opportunities spurred by gradually vanishing trade barriers and expanding world markets have greatly accelerated the food industry's global demand for new products, quality improvements, extended shelf-life, increased productivity, and lower production and distribution costs. Success in the context of these new difficulties will be directly tied to the industry's ability to create more accurate information about customer attitudes and impressions of food items, as well as how they are correctly assessed and utilized. The technique relies heavily on sensory assessment (Difonzo et al., 2019). Sensory evaluation has traditionally been associated with product specialists, and later as a more passive component of the product development team. Currently, the food industry's new challenges need sensory to be more proactive, developing new product concepts based on specific sensory features or separate customer groups distinguishable solely by sensory behavior. However, the survival of sensory assessment as a stand-alone information source is not guaranteed. The sensory evaluation must refine its technique and more specifically explain its responsibilities and function in the food industry.

CHAPTER 7

TYPES OF SENSORY EVALUATION OF FOOD

CONTENTS

7.1. Introduction 156

7.2. Analytical Sensory Evaluation 157

7.3. Discriminative Tests 162

7.4. Types of Discriminative Test 164

7.1. INTRODUCTION

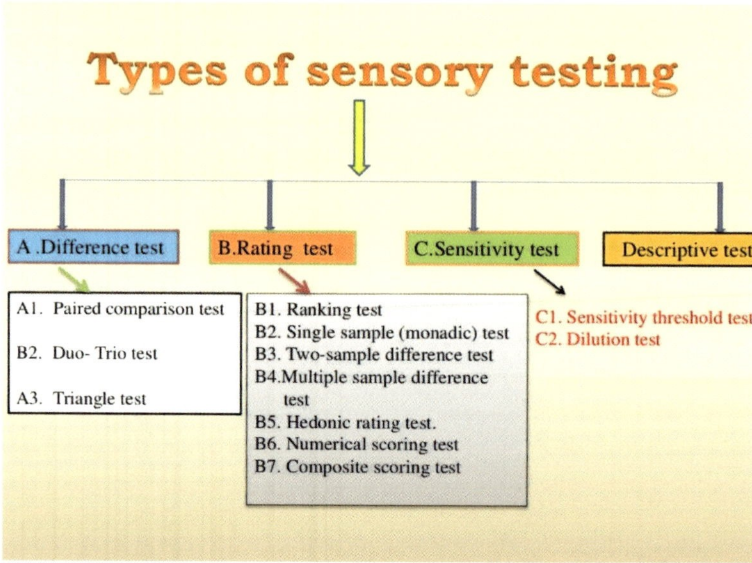

Figure 7.1: Types of sensory evaluation of food.

Source: https://1.bp.blogspot.com/-19eByIPKxJg/WsW3gmogZqI/AAAAAAAAMwI/UPYw2D2WDs8MRvgO-tKt7Wzy-L4FH1yzgCLcBGAs/s1600/sensory%2Btesting.png.

Sensory evaluation (or sensory analysis) is a scientific profession that blends notions of experimental design and statistical analysis to the use of human senses (sight, smell, taste, touch, and hearing) to appraise consumer items. The discipline necessitates panels of human assessors, on whom the commodities are scored, and documenting the replies given by them. By employing statistical approaches to the results it is feasible to generate conclusions and insights about the items under test. Most significant consumer products corporations have divisions dedicated to sensory analysis. Since its start in the 1940s, sensory assessment has evolved into an interesting, dynamic, and perpetually increasing profession that is today recognized as a scientific area in its own right (De Kock & Magano, 2020). Sensory experts are typically presented with issues that need a vast array of skills from various disciplines, including biosciences, psychology, and statistics, and they are routinely pressured to connect with other specialists in these fields. Because of human unpredictability, working with a human as a measuring instrument' is equally tricky.

Today's lifestyle is fundamentally affected; hypermarkets give a huge assortment of gourmet items to consumers. Sensory analysis has become a significant part of food production as food manufacturing enterprises battle for greater shelf space in supermarkets. Sensory evaluation has become a significant component of the design of food items, with considerable developments in the standards for setting up, testing, analyzing, and interpreting sensory data. Furthermore, improvements in electronic device technology have eased the test process. Sensory analysis of food may typically be divided down into three sub-sections:

- Analytical testing (dealing with factual data regarding goods);
- Affective testing (dealing with subjective factors such as preferences);
- Perception (the physiological and psychological components of experience).

7.2. ANALYTICAL SENSORY EVALUATION

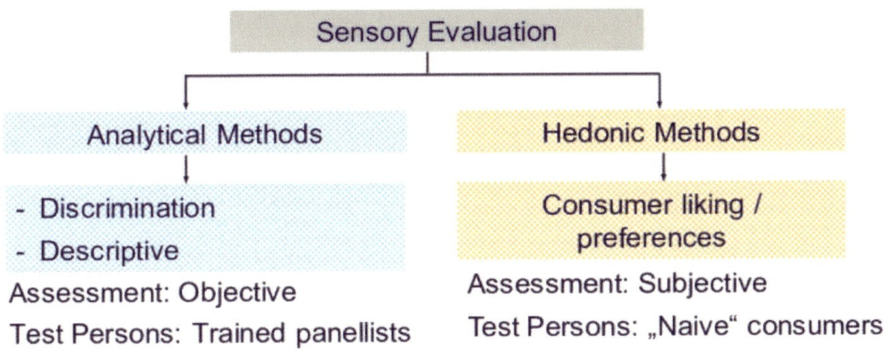

Figure 7.2: Analytical sensory evaluation.

Source: https://www.researchgate.net/profile/Maurizio-Canavari/publication/45229064/figure/fig4/AS:339829511475205@1458033077183/Sensory-methods-are-divided-in-analytical-tests-including-discrimination-and-descriptive.png.

Analytical sensory evaluation is divided into two:

- descriptive sensory evaluation (difference test); and
- discrimination sensory evaluation.

7.2.1. Descriptive Sensory Evaluation

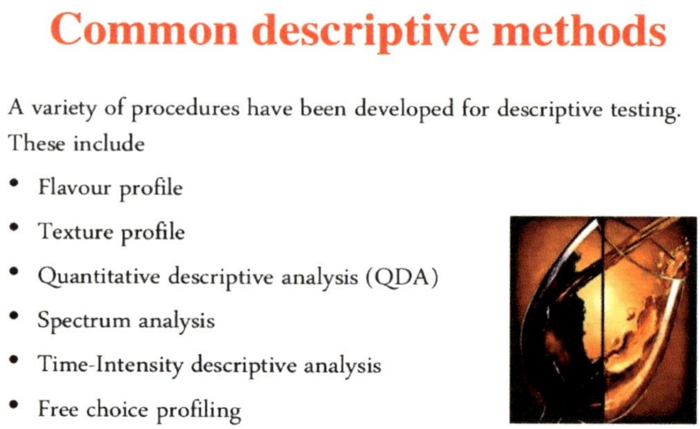

Figure 7.3: Descriptive sensory evaluation.

Source: https://image.slidesharecdn.com/descriptiveanalysisforsensoryevaluation-150628155853-lva1-app6891/85/descriptive-analysis-for-sensory-evaluation-15–320.jpg?cb=1435507438.

Descriptive sensory evaluation is a collection of methods that can be used to differentiate between products based on all of their sensory properties and to arrive at a quantitative description of all observable sensory qualities, not just defects. The term "descriptive sensory evaluation" refers to a collection of methods that can be used. It is of the utmost importance to keep in mind that descriptive sensory assessment is not restricted to only pointing out areas that may be improved (Daher et al., 2020).

In contrast to the more traditional means of assessing quality, such as the employment of scorecards, there will not be any "great" or "awful" ratings handed out during the actual exam itself. This is because there are no such ratings. This is because such judgments are not intended to be made as part of the assessment; hence, to do so would be to undermine the evaluation's own goal. It is feasible for a product to have drastically different sensory features when utilizing more traditional techniques, yet the product might still earn the same quality score as long as the product did not include any flaws. This is the case as long as the product was not flawed in any way. Because of this, a review of this sort does not take into account the differences in sensory features that may be important for the acceptance of a product by consumers. In addition, there is not enough infoxrmation provided for a proper understanding of the physical and chemical factors that contribute to sensory qualities.

As a consequence of the techniques that are taken in descriptive sensory evaluation, researchers, those who are responsible for the development of goods, and those who engage in marketing now have access to an extra instrument. In addition, rather than relying on the expertise of a single person, the processes involved in the descriptive sensory assessment are carried out by a group of evaluators working together. This leads to a consensus that is superior to the judgment of a single expert in the sense that it is less susceptible to prejudice and more resilient against it (Chen & Opara, 2013). As a result of this, the consensus is superior. The results that can be acquired from a sensory panel that has been educated to carry out the tasks of an instrument generate findings that are equivalent to the results that can be obtained from instruments.

A dairy product's scent, color, flavor, texture, aftertaste, and even sound are all examples of sensory qualities. Other examples include how the food tastes. All of these aspects may be utilized to separate one product from another one that is currently on the market. Methodologies for descriptive analysis include, but are not limited to, the flavor profile method, the texture profile method, the Quantitative Descriptive Analysis (QDA) technique, the Spectrum method, the quantitative flavor profiling method, and the free-choice profiling approach. Other methodologies for descriptive analysis include the spectrum method, the free-choice profiling approach, and the quantitative flavor profiling method. Every one of these techniques provides a profile of either the taste or the feel of a substance.

In the descriptive analysis, two further ways that may be employed are the spectrum method and the free-choice profiling method. Both of these methods are referred to as "spectrum methods." During the process of its application, any descriptive approach can be segmented into one of three separate stages. The first thing that has to be done to move forward with the process is to assemble a group of highly skilled professionals who will carry out the sensory evaluation. This is a prerequisite for moving on to the next step (Chowdhury & Das, 2015). In the second stage of the process, the building and definition of language, also known as a vocabulary, is carried out to explain the myriad of characteristics of the product that are related to the various senses(s). The very last thing that has to be done is to measure these sensory qualities. Even if each of the aforementioned methodologies is predicated on a distinctive philosophical perspective and methodological approach to the senses, their end purpose is still the same: to train a group of people to recognize and measure the qualities of the senses.

Because the dairy industry has a long history of evaluating the sensory quality of goods, which is tied to a complete understanding of the qualities of dairy products and the variables that contribute to them, it would be reasonable to build upon this experience rather than rebuild the wheel. This is because evaluating the sensory quality of goods is tied to a complete understanding of the qualities of dairy products and the variables that contribute to them. This is because determining the sensory quality of a product requires a comprehensive knowledge of the qualities of dairy products as well as the factors that contribute to the formation of such qualities. The product category that is going to be evaluated, as well as the estimated amount of time and financial commitment that is going to be allotted to the program of product testing, will determine not only the descriptive method that will be utilized but also the product category that will be evaluated.

Options include the QDATM method, which is based on prior experience and seeks to precisely define and reference each sensory characteristic throughout many meetings with assessors, and the Spectrum™ method, which is based on prior experience and seeks to precisely define and reference each sensory characteristic throughout many meetings with assessors while taking into account time and financial constraints. Both of these methods seek to precisely define and reference each sensory characteristic throughout many meetings with assessors. Throughout several discussions with raters, both of these approaches make an effort to accurately identify and refer to each sensory attribute. Both of these strategies incorporate the utilization of pertinent experience (Bodyfelt et al., 2008). The latter method also enables one to compare the results that were received from a range of panels, laboratories, and different times in time. When using descriptive sensory evaluation, in general, the panel of assessors needs to be able to recognize a wide variety of sensory characteristics that are present in a product, come to an agreement on how these characteristics are perceived and labeled, and be able to score the intensity of each characteristic consistently using a line scale.

In addition to these requirements, the panel also needs to be able to recognize a wide variety of sensory characteristics that are present in a product. In addition to this, the panel must be able to reach a consensus on how these qualities are identified and categorized. Therefore, for the panel to achieve the requisite degree of objectivity and repeatability, the evaluators of the descriptive sensory evaluation need to go through a significant amount of training. For the panel to achieve the requisite level of objectivity, this must first be accomplished. If a panel is going to be used for an undetermined period, it is necessary to have it calibrated and maintained consistently.

The objective of the descriptive testing process is to identify and quantify certain sensory properties. Prospective panelists are required to have the capacity to distinguish between the various types of testing, and their selection is based on their level of proficiency in this area (Chira et al., 2020).

- Evaluation of attributes:

This method, which consists of a string of word phrases arranged either in descending or ascending order of intensity, can be applied to the evaluation of a variety of various aspects (e.g., sweetness, off-flavor, etc.). After that, each point on the scale is assigned a numerical value, with the lowest number indicating that there is no intensity at all and the highest number indicating that there is the maximum amount of intensity that is feasible. An analysis of variance is carried out to determine the degree to which the means of the intensity ratings for each sample differ from one another and to determine whether or not these differences are statistically significant (Coggins, 2007). To calculate ratios and proportions, modularity is necessary. (Magnitude Estimation) The goal of this study is to establish whether or not there is a correlation between the level of sensory intensity and the level of the physical intensity of a given experience.

This program also has the capability of comparing the quality ratings of two or more different products, which is still another application for it. Each participant is provided with a varied pool of numbers from which to pick, in addition to the additional difficulty of assigning numbers by their unique levels of sensory intensity. You are not allowed to add any negative numbers when offering a numerical rating for the initial sample of the subject matter you were given to evaluate. The rating that is given to subsequent samples needs to be a direct reflection of the rating that was given to the sample that came before it. When the process of normalization is complete, the numbers that have been assigned are then incorporated into the statistical analysis. The product's flavor profile and its characteristics. The use of this procedure results in the documentation of detectable odor and flavor components, in addition to sensory characteristics and aftertastes. An intensity level is assigned to each olfactory and gustatory sense note using the scale that is used throughout the process of describing each note. In the vast majority of cases, there will be between four and five people serving on the panel. Each member of the panel conducts their assessments of the item being evaluated, making notes about how they experience the item in terms of its scent, flavor, and finish. After that, they hand their findings over to the moderator of the panel, who then facilitates an open discussion on the topic.

The flavor profile can be affected by a broad variety of factors, such as the ingredients, the manufacturing method, the packaging, the length of time the product has been stored, and so on. An Examination of the Texture Profile This method takes a descriptive approach and is founded on the characteristics of the flavors involved. Utilizing this method, an evaluation of the mechanical, geometrical, lipid, and moisture properties of food can be carried out. The number of people that make up the panel might range anywhere from six to nine different people (Chen et al., 2022). The conclusions of the panel are contrasted and compared with one another using a profile, and the findings are then summarized. In this part, the results of the analyzes are presented in a quantitative format. This technique stipulates that repeated testing is necessary for each product that is being evaluated. The main aspects of the examination include a category rating system that is not organized and at least six trained panelists who are required to be present.

7.3. DISCRIMINATIVE TESTS

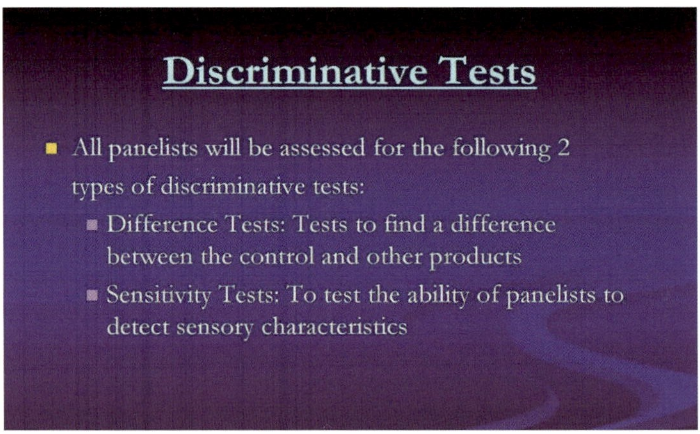

Figure 7.4: Discriminative tests.

Source: https://docplayer.net/docs-images/46/21695818/images/page_19.jpg.

Discrimination testing is a subset of sensory analysis that seeks to answer the question of whether or not two or more distinct characteristics can be distinguished from one another. To provide a reliable means of distinguishing between the various test items, the examination makes use of a group of evaluators (panelists) whose levels of expertise are correlated with the difficulty of the examination (Castro et al., 2013). Even if these tests can be beneficial, it is required to apply a different way of study together with a panel of experts that have obtained further training to explain and analyze the discrepancies. Tests of discrimination can be used by specialists in the senses to assess whether or not two samples can be perceptibly distinguished from one another.

As a consequence of this, it is feasible for two samples to have a chemistry that is remarkably distinct from one another compared to that of the other sample. By selecting this option, the creators of a product are provided with the ability to make modifications to the product without calling the attention of the final consumer to the fact that the modification has been done. Many food producers, such as huge ice cream firms, hunt for cheaper vanilla flavors to utilize in their goods as a substitute for vanilla because vanilla has such a high price tag. The objective of the seller is to achieve a state in which the customer is completely oblivious to any differences that exist between the two choices. If a discrimination test was carried out appropriately and had sufficient power to indicate that the two ice cream recipe formulations were not regarded as dissimilar from one another, then the corporation would be in a better position to make the changeover with a lower level of risk.

By taking a look at this example, one can observe a wonderful demonstration of how sensory discrimination testing works in practice. When the processor alters the processing that they feel will not affect the sensory qualities of the product, discrimination testing may be employed. This type of testing may also be used when the product is being evaluated by a consumer. The discriminating test does not provide any evidence to suggest that the null hypothesis ought to be rejected under any of these conditions. A corporation can create a completely new product by making slight adjustments to an existing one. In the case that a "new and improved" version is available, it is feasible to perform the discrimination test to evaluate whether or not the two are different formulations are seen as being distinct (Yağcı & Göğüş, 2008). This may be done to decide whether or not the discrimination test can be used. In light of the facts presented above, the purpose of the discriminating test is to demonstrate that the null hypothesis is false.

If the results show that the two different formulations are considerably different from one another, the sensory scientist has a responsibility to experiment to discover whether or not the "new" formulation is more effective than the other formulation. The formulation is widely acknowledged to be a move in the right direction, and this understanding is widespread. The discrimination tests are made worthless when there are big and clear variations between the samples. If preliminary bench testing suggests that there will be a detectable difference between the two samples for each member of the panel, then it is not suggested to apply these techniques. Rather, other approaches should be utilized. Scaling methods might help establish the true extent of the difference that exists between the samples in circumstances such as these.

In other words, the cases in which the sample variances are modest are the ones in which discrimination testing has the best chance of being effective. Errors of type II, on the other hand, are prone to occur as a result of even the slightest of deviations from the norm. Discrimination tests are typically used in instances when there are only two samples available for testing. The results of these tests can be used to draw comparisons between more than two distinct aspects; nevertheless, they are laborious, statistically unreliable, and ineffective. Grading systems and scales are two examples of strategies that perform well in a variety of contexts. When it comes to discriminating tests, one has a lot of alternatives from which to choose. Some examples of these tests are the triangle and duo tests, the paired comparison tests, and the n alternative forced-choice tests.

7.4. TYPES OF DISCRIMINATIVE TEST

Within this general class are a variety of specific methods, e.g.,

- paired comparison test;
- duo-trio test;
- triangle test;
- multiple sample test;
- other test methods such as dual and multiple standard tests.

Additional testing processes such as dual and multiple standard tests are two examples of those that may be utilized. A basic inquiry serves as the driving force behind every one of these approaches.

"Does every item has its unique characteristic?" As you can see, the solution to this riddle may have far-reaching repercussions if it is implemented successfully. For management to have faith in the findings of a discriminating test, it is essential that every part of the test, beginning with its design and continuing through its preparation, processing, and interpretation of the data, be thoroughly studied (Yang & Boyle, 2016). This includes all aspects of the test. The findings of the discrimination test will only have any credibility in the eyes of management once this step has been completed.

7.4.1. Paired Comparison Test

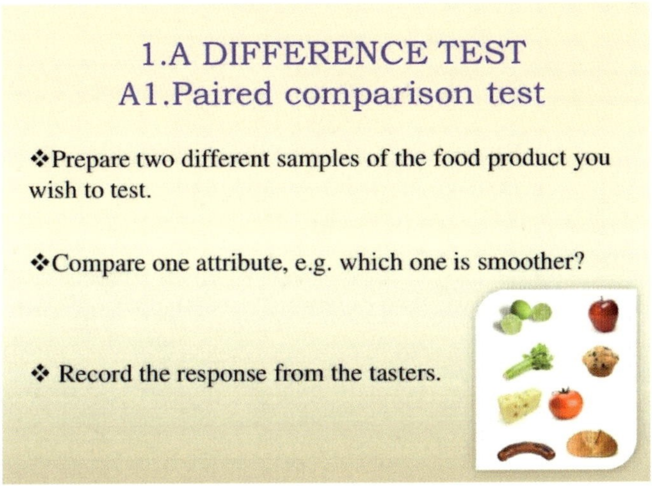

Figure 7.5: Paired comparison test.

Source: https://image.slidesharecdn.com/scorecard-141031091151-conversion-gate01/85/sensory-evaluation-of-food-products-7-320.jpg?cb=1414746955.

In the process of evaluating food and drink, paired comparisons are put to the test so that discrimination may be determined. In addition to this, it has been utilized effectively to determine flavor thresholds for fundamental solutions. In a test when two items are compared side by side, it is the responsibility of the panelist to determine which item is superior in terms of a certain attribute, such as sweetness, skinniness, or softness. This procedure is sometimes referred to as a directed paired comparison test, which is a frequent word. During the "directed" portion of the exam, the panelist is instructed to follow a certain structure for the paired test. The organization and execution of paired comparison tests are not complicated at all. They are questioned about whether or not they observe any distinction between the objects that are coded (AA, BB, AB, and BA) and those that are not coded (C, D, F, G, and H). It has been demonstrated that demanding a "different" reaction to each scenario leads to significantly improved outcomes.

The A-not-A approach is yet another potential variation of the paired test that may be used. During the assessment, the participant will first be presented with a single sample, which will later be exchanged for a second sample. After making this judgment, the subject is responsible for deciding if the goods are the same or distinct from one another (or different). When adopting this one-of-a-kind testing technique, it is feasible that findings might be influenced by elements other than those being examined, such as differences in color. Researchers can determine which of two samples has a larger concentration of an under examination attribute by employing paired comparison tests, or which of the two samples is more desirable in general by determining which of the two samples is more favorable (Warnakulasooriya et al., 2020). The second usage of the term is seen as an acceptableness check due to its frequent use. The panelists must have a complete comprehension of this common attribute difference test.

7.4.2. Due–Trio Test

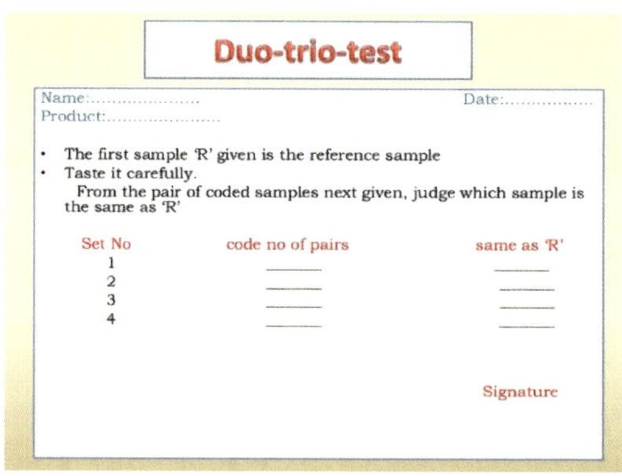

Duo-trio-test

Name:.................... Date:................

Product:.....................

- The first sample 'R' given is the reference sample
- Taste it carefully.

From the pair of coded samples next given, judge which sample is the same as 'R'

Set No	code no of pairs	same as 'R'
1	______	______
2	______	______
3	______	______
4	______	______

Signature

Figure 7.6: The Due–Trio test.

Source: https://image.slidesharecdn.com/scorecard-141031091151-conversion-gate01/85/sensory-evaluation-of-food-products-10–320.jpg?cb=1414746955.

The Due–Trio test is useful for evaluating whether or not a product has significant gustatory, olfactory, or other sensory impacts. As a result of this test, the sensitivity is significantly lowered. It is helpful for jobs like quality control and judge selection since it allows for greater discrimination than other methods. The due-trio test, along with the other two ways of product testing, has a probability of chance that is comparable to that of the other two procedures. The due—threefold, constant—reference test approach is the greatest choice for making comparisons with items that are currently on the market.

Because the chance probability correlation with the test is just one-third, advocates of the "three product" (or "triangle") test believe that its claim of greater sensitivity is well-founded. This is because the chance probability correlation with the test is only one-third. The triangle test is a more difficult examination, and in order to correctly complete it, it is necessary to recall not only two sensory aspects of two distinct items but also the judgment of a third object. This test can be seen as a mixture of three distinct paired tests, namely (A and B to C), and (A-B, A-C, B-C). It is not uncommon for the triangle test to be impossible to be carried out because of goods that are difficult to remove from the palate, as well as products that have physiological effects (distilled beverages).

The triangle test is used to examine whether or not there are discernible variations in terms of a person's sensory experiences when comparing two different things (Tsai et al., 2021). It is especially useful in situations in which changes in the production process might have led to variations in the product. The panelists investigate the three samples to find out which one stands out from the rest. Because the duo-trio test, the paired comparison test, and the triangle test are all reliant on chance, the triangle test is more effective than the duo-trio test and the paired comparison test combined. If the panelists are thrown off by the fact that there are three samples to evaluate rather than two, then using a test with three samples might not be the most effective choice.

7.4.3. Multiple Sample Test

Tests that involve more than three different stimuli are referred to as multiple sample tests. It is possible for there to be an equal quantity of symmetrical and asymmetrical stimuli present (called asymmetrical). For a judge to determine whether or not there is a significant difference between two samples, the samples themselves need to be divided into two groups of samples that may be compared to one another. When utilizing these questions in guided examinations, judges need to differentiate between groups that are more or less intensive, or that have a certain quality. It has been discovered that including more than three distinct stimuli in a different test design is not very useful. Justifying the restriction is the fact that increasing the number of stimuli causes the subject to experience a rise in both the psychological complexity of their responses as well as a commensurate increase in their bodily tiredness. In addition to this, a significant number of samples have to be collected, and the observer needs to invest a great deal of time before arriving at any conclusions (Ayzeli et al., 2020). Visual discrimination appears to be the most appropriate usage for these tests since the judge does not need to rely on memory and is nearly free of weariness. This makes visual discrimination the most acceptable use.

7.4.4. Dual Standard Test

To achieve this goal, a technique of quality control that makes use of two different standards was developed. The participant has presented four separate objects, two of which are identified as "references A and B," and the other two of which are coded. Comparisons need to be made between the coded product and the reference product. It's possible that the categories in the two sources hint at problems with quality control, or that they signal the fact that production and the product themselves are both at and very close to their maximum levels.

7.4.5. Multiple Standards Test

When analyzing scents, this test was developed so that a non-uniform standard may be contrasted with an uncertain variable. Within the survey, respondents are provided with two criteria to choose from. They are encouraged to choose the criterion that is the most different from the other criterion. The number of samples that are collected in addition to the total number of samples is indicative of the potential for accurately determining the identity of the unknown.

During the paired test, some sensory specialists discovered that the due-trio was more sensitive than the triangle, whilst others discovered that the triangle was more sensitive than the due-trio. Based on the levels of sensitivity that they possess, the many tests that may be utilized to identify differences can be arranged into the following categories: paired, dual standard, duo-trio, triangle, and multiple standards (Tuorila & Monteleone, 2009). Recent research has found that there is no difference in the sensitivity of the various discrimination tests.

7.4.6. The Triangle Test

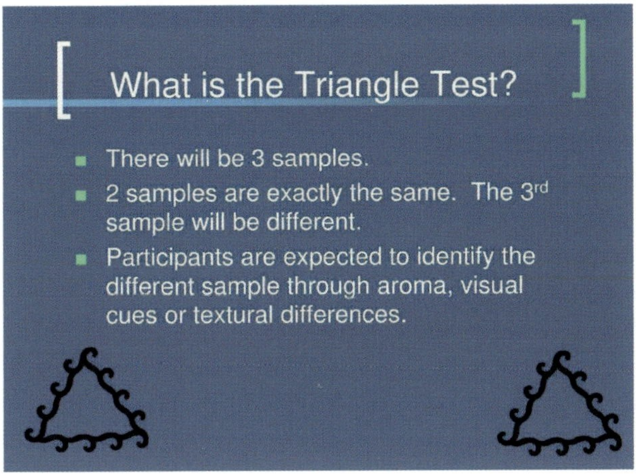

Figure 7.7: The triangle test.

Source: https://slideplayer.com/slide/12991318/79/images/2/What+is+the+Triangle+Test.jpg.

The triangle test is a discriminative approach that can be used in a wide range of settings and functions, making it one of the most popular procedures in the field of sensory research. Its uses may be: to determine whether or if two items are distinguishable from one another in a meaningful way, decide among the individuals who should serve on a certain assessment panel

The procedure of determining if the production method or the components of a product have been altered considerably and then reporting those findings. When conducting a triangle test, each panelist is provided with three samples: one that is distinct from the others, and two that are analogous to it. After a panelist has been provided with all three samples, they should be asked to evaluate them in the order in which they were presented to them (left to right). It is recommended that each member of the panel be allocated one of the six different order combinations that are conceivable. Six different order combinations are available for samples A and B, and they are as follows: AAB, ABA, BAA, and BBA. It is the responsibility of each panelist to make a note of his or her reaction to the one-of-a-kind sample. The triangle test is used to determine whether or not there are discernible differences in sensory characteristics when comparing two different objects. It is especially useful in situations when little adjustments made during manufacturing might have led to significant differences in the final product. The panelists analyze each of the three samples to determine which one stands out from the rest (Short et al., 2021. Because the duo-trio test, the paired comparison test, and the triangle test are all dependent on chance, the triangle test is more effective than the duo-trio test and the paired comparison test combined. If there is a significant amount of carryover flavor from sample to sample, and the panelists are perplexed by the fact that there are three samples to evaluate rather than two, then this test may not be the best option available.

7.4.7. Rating Tests

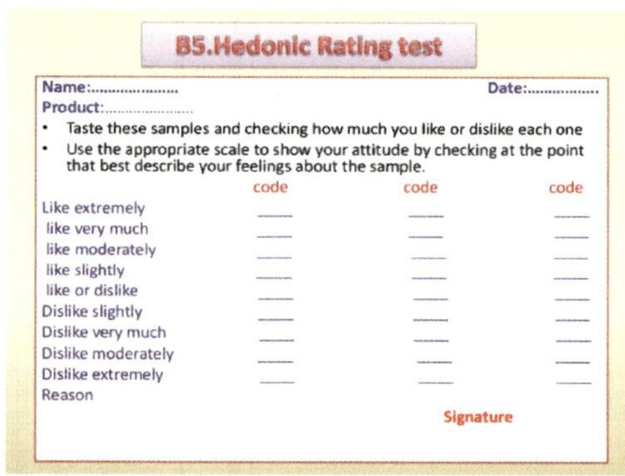

B5.Hedonic Rating test

Name:.................... Date:................

Product:.....................

- Taste these samples and checking how much you like or dislike each one
- Use the appropriate scale to show your attitude by checking at the point that best describe your feelings about the sample.

	code	code	code
Like extremely	____	____	____
like very much	____	____	____
like moderately	____	____	____
like slightly	____	____	____
like or dislike	____	____	____
Dislike slightly	____	____	____
Dislike very much	____	____	____
Dislike moderately	____	____	____
Dislike extremely	____	____	____

Reason

Signature

Figure 7.8: Rating tests.

Source: https://image.slidesharecdn.com/scorecard-141031091151-conversion-gate01/85/sensory-evaluation-of-food-products-22–320.jpg?cb=1414746955.

The panelists don't need to complete further tests; rather, they are required to assess a large number of samples under certain criteria while also completing rating exams. The following are some examples of the many different kinds of tests that are used for assessment:

The panelists do a ranking test, in which they assign points to various attributes of samples and then score those samples overall. The outcomes of the ranking test are discussed in front of the entire panel. It is possible to use this kind of test to compare the quality of a variety of different things.

Panelists are asked to decide whether or not a single feature exists, as well as to what degree it does exist, as part of a sample test that is also known as a monadic test. One sample at a time is all that can be evaluated using this method. There is a possibility that more than one sample may be examined; however, the panelists will not compare the results obtained from the various samples (Sutrisna et al., 2017). A panelist is tasked with determining the degree of similarity or dissimilarity between two sets of samples that have been shown to them in pairs. These samples have been supplied to the panelists individually. At the very least, four separate sample pairings are evaluated throughout a typical run of this test.

One of the samples, which is designated as the "standard" for this particular examination, is distributed to each panelist in an amount ranging from three to six copies. During this test, the results of a few different samples will each be compared to a single "standard." They have to assess the degree to which the standard and the other samples resemble one another and then determine how comparable the other samples are to the standard. Participants in a hedonic rating test are given a variety of delicacies to try, and they are awarded points based on how much they like or detest the meals that they are rating. The hedonic rating test is one type of taste test that may be done. The practice of asking tasters to remark on the sensory aspects of a product or to assess each sample based on the various sensory components they enjoy or dislike as a matter of normal procedure has become commonplace. Tasters are usually asked to comment on how the product makes them feel after trying samples of it.

Participants in the examination are required to assign numbers to each sample by the scale, and each sample has been given a numerical value that corresponds to a predetermined level of quality on the scale. Those who participate in the exam are given a numerical score system as a label. The participants in this test are given a sample, and they are tasked with assigning points to a variety of characteristics using the proper numerical scale. You have the option of having this test performed either before or after the other tests. A wide range of goals might be served by conducting analyzes on a single sample or a large number of samples.

7.4.8. Affective Test

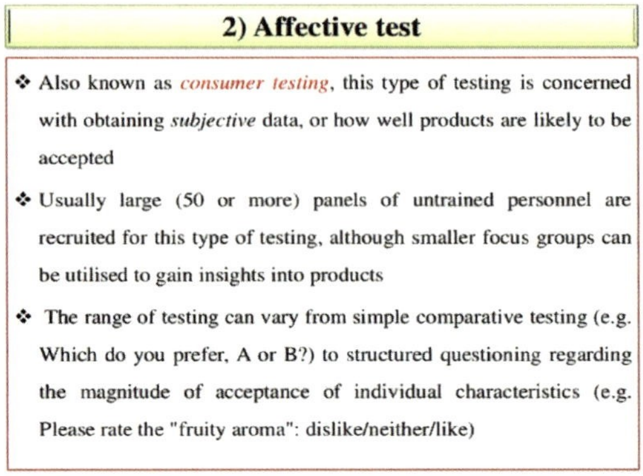

Figure 7.9: Affective tests.

Source: https://image.slidesharecdn.com/sensoryevaluationofessentialoil-finalpptby-shivanandm-r-130107001431-phpapp01/85/sensory-evaluation-of-essential-oil-final-ppt-by-shivanand-mr-29–320.jpg?cb=1357517771.

These are tests that evaluate subjective attitudes like a person's approval of a product or their preference for one over another. When conducting effective tests, the participant is tasked with selecting, rating, or scoring samples to demonstrate their liking or acceptance of a certain stimulus. The paired-preference test and the 9-point hedonic scale are two examples of prominent tests that fall into this category. Respondents are often customers who are chosen based on whether or not they already use the product or have plans to do so in the future. In laboratory settings, consumer demographics are frequently ignored in favor of more easily accessible respondents (e.g., workers) whose preference and acceptance behavior properly coincide with those of the intended customer group.

Acceptance tests of a laboratory kind can be carried out with somewhere between 25 and 50 participants. When doing field research using the population of interest, the minimum number of participants is typically increased by 75–200 or more (Schiling & Pham, 2012). Because of their experience and the possibility that they would provide a biased answer, technical, marketing, and administrative professionals who are connected with a particular product should not be employed in affective testing as a general rule.

A paired preference test, a preference ranking test, and the hedonic test technique are examples of affective approaches that are utilized often. The testing procedure ought to be straightforward to comprehend so that the customers who will be serving on the panel will know how to react. A test of paired preferences. After it has been determined that there is a meaningful distinction between the two wines, a preference test may then be carried out. It is helpful to determine, for example, whether wine mix or yeast fermentation is chosen based on the results of this analysis. It is essential to keep in mind that a preference test should be carried out on its own and after a difference test has been completed. Although it may be tempting to mix the two, you should resist the urge to do so because doing so might lead to inaccurate findings. Before tasting the wine, the tasters need to think about (and potentially discuss) the ideal wine style that is required, since this will play a role in the determination of preference. The selection of the preferred wine should not be based on one's personal preferences but rather on the wine that is most suitable to the desired wine style.

The purpose of these tests is to provide an answer to the issue of which product is favored or how much liking there is for the product. Participants in these testing should ideally be people who have already used the product in question. There is no requirement for training for the panel. There ought to be more than thirty people participating on the panel at the very least. There are a few different approaches to determining one's preferences and level of acceptance. A common one is a hedonic scale, which has 9 points. The paired preference test is organized in the same way as the paired comparison approach that is used for the discrimination test. When deciding which product to purchase, two different samples are compared to one another. To identify which of the two products is favored more, it is necessary to collect a big number of replies that are comparable to one another. A statistical table may be used to help calculate the minimum number of replies that are required to assess whether or not the preference is significant. This minimal number is dependent on the total number of responses that were collected.

The examples of affective tests are discussed below.

7.4.9. Paired Preference Test

Customers of potato chips participated in a preference test to evaluate their eating habits and selections. In the experiment, it appears that the "placebo" chips and the "target" chips were the same things, hence they may be said to be equivalent. This data was subjected to a Signal Detection analysis before being used. Customers who preferred the first type of chip in the target pair were given one distribution based on their "magnitude of preference," whereas customers who preferred the second type of chip were given a distribution based on their "second magnitude of preference." The second set of distributions was built to accommodate the two groups of placebos that were also presented to the clients (Shinde et al., 2016). The d′ value of each chip shows the difference between the preference distribution of the target pair (Signal + Noise) and its equivalent placebo distribution.

This difference was determined by employing a signal detection paradigm, and it was recorded on each chip (Noise). A significant advantage of this study is that the values of D are unaffected by participants' preferences for the placebo pair. This benefit is not taken into consideration by the standard method for estimating d' values, which is based on the 2-AC test. The importance of preference testing cannot be overstated when it comes to determining customer approval. One of the drawbacks of this strategy is that it is susceptible to a phenomenon known as "response bias." This is a situation in which customers select goods that at first glance appear to be distinct from one another, but are, in reality, quite comparable to one another. It is feasible to explore the influence of using comparable products as a control on the findings for the two separate things that are being evaluated if identical items are used as a control. Several different statistical approaches and models might be utilized to find a solution to this issue. In paired preference tests, customers are given a negative answer if they are unable to pick between two different foods. If they do not have a preference, however, they are given a choice between both dishes.

When conducting preference tests for meals, researchers ask participants to select one of two alternatives and indicate whether or not there is a discernible difference in the likelihood of purchasing each choice. On the other hand, the first one forecasts decision-making, and the second one forecasts purchasing behavior. It is necessary to monitor this behavior for several months to demonstrate that the testing is capable of properly predicting the behavior of customers. Participants may also be offered the opportunity to consume some of the meals that were utilized in the study in the comfort of their own homes (Sun et al., 2021). Customers of potato chips were encouraged to sample a wide selection of products, each of which featured a distinctive flavor and outward look. The initial group of participants in this research project (with a total of 109 individuals), were given a conventional paired preference test. In a second group with a total of 108 participants, those individuals were questioned on various aspects of their buying habits. A third group of students, with a total of 101 participants, was given both exams.

Two rows of straightforward plastic snack bags with the test chips hid the act of consumption from the view of the researchers experimenting. A "Take Away" preference is a want to take one bag of each flavor of chips, whereas a "Take Away" preference is a desire to take two bags of the same variety of chips (rejection). Customers in the first group had a 48% chance of properly predicting what had been removed, whereas customers in the second group had just a 45% chance of accurately predicting what had been taken away (third group). In terms of 'purchasing' preferences, there was a relationship between the second group, which included 56%, and the third group, which comprised 60% (third group). The outcomes were different depending on whether the liking tests were carried out on their own or in combination with the purchasing behavior surveys. More than sixty percent of buyers across all categories choose the more reliable option, which was to buy one bag of each product. The predictability of 'Take Away' decisions was restricted in paired preference experiments of liking and buying. These studies were conducted. .... The "Take Away" selections that a client makes are an excellent place to start in the process of getting to know him or her better.

7.5. 9-Point Hedonic Scale

Hedonic Scale	Score
Like extremely	7
Like very much	6
Like	5
Dislike slightly	4
Dislike moderately	3
Dislike very much	2
Dislike extremely	1

Figure 7.10: The hedonic nine-point scale.

Source: https://www.researchgate.net/profile/Radius-Tanone/publication/336049007/figure/tbl1/AS:807186330550273@1569459626919/Hedonic-Rating-Scale-and-Score-13.png.

The hedonic nine-point scale is the approach that is utilized the majority of the time when determining whether or not a meal is pleasurable. To investigate the preferences of military members, the Quartermaster Food and Container Institute of the United States Armed Forces developed a nutritional preference measure. The scale is used to determine whether or not the cosmetics, personal care items, and other household goods may be consumed safely by eating or drinking them.

The hedonic scale was initially developed as a result of extensive research carried out at the University of Chicago and the Quartermaster. The research of Jones, Peryam, and Thurstone showed that lengthier scales that included up to nine intervals were superior to shorter scales in terms of their discriminatory power. It was also argued that a scale consisting of eleven intervals would be far more effective3. The nine-point variety was chosen as the standard by the Quartermaster in the end because it took up less space on the type of paper that was utilized for the printing of the ballots (Serventi et al., 2022). As a result of this, it was feasible to position each point on the scale with a similar degree of psychological distance from the point that came before it 3. For instance, this equal-interval criterion mandates that successive integer values (1, 2, 3, etc.) be assigned to scale points, and that average acceptability changes be computed utilizing parametric statistics, as was just demonstrated.

The validity, reliability, and discriminatory capacities of the scale have been shown by a significant number of field and laboratory investigations on food acceptance as well as large-scale surveys of people's preferences about food. The original hedonic scale, which consisted of nine points, has been refined multiple times. It was shown that a scale with nine "child-friendly" verbal anchors, ranging from "superb" to "poor," performed significantly better with children aged 5 to 10 than one with "smiley" visuals. These anchors ranged from "superb" to "bad." These anchors spanned the full range of quality, from "superb" to "awful." The original scoring system, which consisted of nine points, has been adopted by other languages. It is essential to keep in mind that the original characteristics of the 9-point scale do not always match the scale's translations in any way. To choose the verbal anchors and check the reliability, validity, and discriminating capacities of the proposed scale, research similar to that carried out by Peryam and colleagues would be an appropriate choice (Silva et al., 2020). At the very least, some research has to be conducted to guarantee that the ranking order of the verbal anchors can be determined with complete assurance among the population that will serve as the basis for the examination. The nine-point hedonic scale can be shown in a variety of ways, including vertically as opposed to horizontally, and for the latter, with "like strongly" or "dislike excessively" on the left.

Alternatively, the scale can be depicted in a vertical format. According to the research that was conducted in 1957 by Peryam and Pilgrim3, the changes made to the experiment had "no meaningful influence" on the results. According to the findings of the study that Peryam and Kroll conducted at the Quartermaster, alterations to the display of the scale can affect how the average answer is determined. It is vital to maintain a consistent presentation of the scale throughout the whole inquiry to facilitate comparisons with previous research or an absolute standard (e.g., a score of 7.0 on a scale with 9 points).

CHAPTER 8

HOW DO OFF-FLAVORS OCCUR?

CONTENTS

8.1. Introduction to Off-Flavors 180

8.2. The Smell of Chlorophenol May Be Found in Iodine, Plastic, and Vinyl 185

8.3. Freshly Mowed Grass With a Strong Aroma 188

8.4. Husky/Grainy 189

8.5. An Introduction to Hygiene 196

8.6. Fermentation's Undesirable Flavors and Their Management 196

8.7. The Most Commonly Observed Emotions 198

8.8. Discussion 198

8.9 Summary of Off-Flavors 201

8.1. INTRODUCTION TO OFF-FLAVORS

Customers are more worried about the environmental and human health impacts of traditional farming practices in recent years. Many agribusinesses have made initiatives to differentiate their goods to overcome these hurdles and acquire new customers in highly competitive markets. The rise in demand for ecologically friendly products has sparked a product boom in industrialized countries. Schaufele and Hamm studied the preferences and willingness to pay extra for environmentally friendly wines. The production and marketing of wine with sustainability qualities is one possible strategy for quality differentiation, particularly for regional and organic wines, according to the research. According to actual evidence from wine experts, the image of organic goods has a higher impact on customer perception than the underlying features (Rousta et al., 2021). When comparing the two, this was proven to be the case. The blurring effect, often known as the halo effect, can alter the perception of a product. This has a favorable effect on organic and origin-labeled items, but a negative effect on conventional goods. A blind taste test conducted by Pagliarini found that participants couldn't tell the difference between organic and conventional wines. Customers were more inclined to choose organic wine when they had more information at their disposal.

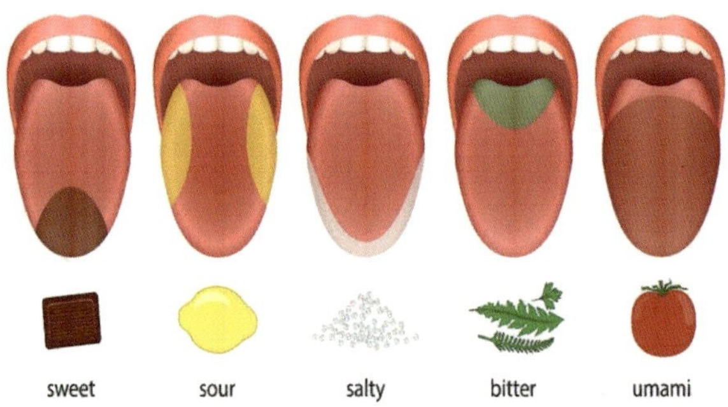

Figure 8.1: Flavor Images.

Source: https://www.yummylixlollipops.com/food-for-thought-the-5-basic-taste-categories/.

This study suggests that customers' greater willingness to pay extra for organic wine is due to their differing attitudes and interests in environmental concerns. Educating consumers on the possible dangers of organic wine to their health, safety, and the environment was also proven to improve their perceptions of the product. It was found by Apaolaza's research that organic wine has a more refined, strong and flowery smell as well as a smoother texture, making it more enjoyable to drink (Pu et al., 2020). There is a psychological or emotional underpinning for this occurrence, which is countered by an increase in sensory ratings and perceived wellness. These studies reveal that organic wines have greater quality than conventional wines, and they also show that the halo effect of storytelling is present regardless of whether or not a customer is aware of organic goods or how they feel about them.

They also provide consumers with varying levels of wine knowledge and those who have tasted wines with no discernible off-flavors have responded to the study thus far. In our opinion, there is no evidence to suggest that highly trained wine tasters favor aromatically flawed organic wines. Some experts believe that this is because highly trained wine tasters are more capable of discerning the distinctions between wines made with conventional and organic processes. No research has been done on this topic. That is why we wanted to find out if people appreciate organic wines if they will pay for them, and what characteristics they have in terms of taste. Recent decades have seen ground-breaking research on the flavors and aromas of a wide range of drinks. It is possible that alcohol use still accounts for the majority of the discrepancy in flavonoid and aromatic component characterization, even though these chemicals have been extensively studied. To provide a consistent taste and experience in the mouth, most brewers employ rigorous quality control measures. An off-flavor may be introduced into beer, which is the most well-known example of this sort of brewing. Regardless of whether you're making a commercial or a personal batch of beer, you'll need a high degree of skill and process optimization to get the best results.

Bitterness, alcohol concentration, and other sensory characteristics can all have an impact on a product's market value. The length of the fermentation process and the pace at which the yeast is pitched are both critical, according to the research done thus far. Beer off-flavors can be formed as a result of a beer's fermentation characteristics. Controlling off-flavors necessitates a large amount of high-quality yeast (Peinado et al., 2014). The water used in the brewing process might vary in quality from place to region, which can result in varied tastes and off-flavors. It is possible to produce more than 10 different unpleasant flavors in beer throughout the brewing process. Off-flavors are currently documented in fragmentary literature; nonetheless, to the best of our knowledge, no comprehensive data exist to identify all off-flavors.

In addition, the brewer must be able to distinguish between good and bad flavors. To be a brewer, one must be able to discern the flavors. Chemical evidence is the sole evidence of off-flavors that can be identified in the current corpus of study; sensory evidence is rare. When off-flavors are discovered in a brewery, the question of where they come from is commonly discussed. Studies linking all off-flavors to their sensory features and therapies are rare. By reading the material on this page, you will have a better understanding of all off-flavors in beer and their causes and possible remedies. Common tongue's sensory qualities help the brewer find and remove unwanted tastes from the end product. Many therapies exist, but not all of them have been proven to be effective by scientific research. Yeast can also be utilized to mask disagreeable flavors, as was previously stated.

The flavor is a result of the interaction between the olfactory and gustatory senses. Some of these feelings include a chiliness or burning sensation, as well as pungentness and discomfort when food is sweet or salty. Assuming that there are just four to seven or eleven fundamental flavors is more of a convenience than an accurate depiction of the truth, even though sweetness, bitterness and saltiness are all essential components of a taste. Different elements make up the flavor of a beer. In most cases, the concentrations are so low that they are not detectable by organoleptic means in the first place. 'off-taste' describes the flavor of a beer that has been improved but still has a flavor that is unpalatable to drinkers. An increase in the number of regular taste components or the presence of one or more chemicals that are not ordinarily found in beer might generate off-tastes.

A variety of tests are used by scientists to gauge the beer's bitterness. Bitterness may be induced by tannins and other chemicals, therefore measuring the concentration of "isohumulones" or "isoids" can only go so far. A sugar analysis should be adequate to determine the degree of sweetness in a beer because it is unlawful for brewers to use saccharin or cyclamate. Salinity isn't a defining trait of beer, thus brewers are unlikely to pay attention to it as a flavor component. Beverage acidity (or pH) may be used to gauge "sourness," while gas chromatography can be used to identify individual acids. A brewer, on the other hand, is not required to utilize any sort of gas chromatography instrument (Perera et al., 2014). Thus, he must undergo sensory testing, which entails the evaluation of the beer's aroma, taste, and overall quality. Hircine, Goaty, and Caprylic were all included in Linnaeus' first systematic olfactory categorization in 1758, along with the scent class. Caprylic acid shares a fragrance note with hexanoic, octanoic, and decanoic acids, which is a more accurate descriptor. Caproic, caprylic, and capric all get their names from the Latin word for goat, caper. There are caproic acid, caproic fatty acids, and caproic acid. Because octanoic acid is the most abundant free fatty acid in beer, it follows that it has a significant impact on the flavor.

Lager yeast *S. uvarum* was used instead of ale yeasts to brew these beers. A larger concentration of octanoic acid and decanoic acid was produced by S. cerevisiae, together with a more soapy and greasy taste effect. The chemical term for caprylic acid is caprylyl acetate, and it has this flavor. When it comes to brewers, the yeasty flavor may be recognized because of its association with a certain yeast strain or type. During the primary fermentation of beer wort, malodorous sulfur compounds, such as hydrogen sulfide and mercaptans, are also formed as intermediates or end products of yeast metabolism, to a degree that depends on the yeast genetic makeup, and the prevalent fermentation circumstances. This is because the yeast's capacity to convert carbohydrates into a form that can be utilized for the production of foul-smelling sulfur compounds is essential.

Since these chemicals are volatile, most of them will be emitted as carbon dioxide during fermentation, but even so, small quantities of these compounds may remain in the beer and harm its aroma and flavor. The maximum concentration of H2S that may be discovered in wort throughout the fermentation process is between 0 and 2 ppm. Even at 0 ppm, it has been identified in completed beers from throughout the world (ppm). Although finished beer normally has a concentration between 0 and 1 parts per million (ppm), studies have shown that concentrations as low as 0 ppm have been achieved." A reading of 0 fl ppm is not out of the ordinary. Fruity aromas are created during the fermenting process. Several factors contribute to a beer's flavor, and esters are one of them. Beer's flavor is boosted by esters, which are flavoring agents that enhance the flavor of the beer.

Sensory smell thresholds for ethyl acetate range from 0.2 to 15–20 ppm, but those for isoamyl acetate are 0.2 to 0.1 ppm. Ester molecules are created by yeast cells, and their length enables them to permeate the liquid around them. Flavors with fruity or metallic notes aren't the only ones to avoid. A molecule called 3-methyl-2-buten-l-thiol imparts a "light-struck" taste to beer that has been exposed to sunshine. In the photolysis of hop iso-acids, such as iso-humulone, iso-cohumulone, and isadhumulone, this chemical is formed (Predieri et al., 2021). In certain cases, brewers' taste panels have detected a second off-flavor with characteristics that have been described as "onion-like." Many references to off-flavors resembling onions may be found in the brewing literature.

There have been studies on the influence of different flavors on beer as a result of this. On the other side, there has been no investigation into every possible taste that may be found at any particular moment or location. With today's current research, beer off-flavors may be studied and analyzed in one place by brewers, allowing them to better understand their origins and their sensory characteristics.

The aroma and taste of acetaldehyde evoke visions of juicy green apples, rotting red apples, and newly carved pumpkins. Some of the likely causes are as follows: The spontaneous generation of acetaldehyde throughout the fermentation process is the work of the yeast. In beers with a high alcohol content or when insufficient yeast is pitched, the process of turning it into ethanol alcohol may take longer. This occurs as a result of the yeast's requirement for additional time to complete the fermentation process. Bacteria can also contribute to the green apple flavor. To stay on top of it, here are some tips: Allow the brew to mature for many months to bring forth its full taste and aroma. To complete the fermentation process, the yeast must be able to convert acetaldehyde into ethanol. Make a yeast starter or use high-quality yeast and pay close attention to pitching the correct amount based on the gravity of the wort.

The flavor is peppery and abrasive, with a "hot" aftertaste at the back of the throat and a strong alcohol aroma. A paint thinner, acetone and acetone, as well as acetone and acetone

Most of the bad tastes in alcohol may be traced back to phenolic alcohols like tyrosol, but also to fuel alcohols like propanol, butanol, isobutanol, and isoamyl alcohol. The presence of these alcohols is more noticeable in beers with a lower alcohol content than in those with higher alcohol content, such as barley wines or strong ales, when just a little amount is desired. These alcohols are most often created by fermentation overheating, but they can also be produced by oxidation, which occurs when the beer is left on the trub for an extended period (O'Mahony, 2018). To stay on top of it, here are some tips: While fermentation is occurring, the temperature should not exceed 80 degrees Fahrenheit at all times. To ensure that the beer stays in the fermenter for a sufficient amount of time, it's best to strain the wort of any sediment before adding it to the fermenter. A secondary fermenter can also be used to reduce the length of time the beer is in contact with the trub.

It has a sour, vinegary, tannic, drying, puckering in the tongue, powdery, and metallic flavor that is akin to chewing on grape skin or a tea bag. Some of the likely causes are as follows: Astringency can be influenced by a variety of things. Polyphenols, more often known as tannins, are the primary source of most of these tastes. The husks and skins of grains and fruits contain tannins. To make tannins from grain, the grain must either be left to soak for a lengthy period or undergo a large amount of processing or crushing. Astringent tastes may occur if the pH of the mashed potatoes is above 5.2 to 5.6. Astringent properties can also be caused by over-hopping the final product. Finding a Workaround: Avoid grains that have been over-milled. However, the grain should not be pulverized or shredded. When sparging, pay attention to the temperature and volume of water used. Remember to remove the grains from the water just before it begins to boil while steeping grains. A fermenter or warm water can be used instead of boiling the fruits in the wort for 15–30 minutes. Make certain that the hops used are appropriate for the beer being produced by checking the type and amount of hops used.

8.2. THE SMELL OF CHLOROPHENOL MAY BE FOUND IN IODINE, PLASTIC, AND VINYL

Aromas that resemble medical supplies or plastic are most commonly caused by using chlorinated tap water, whether for cleaning or making beer. Chlorine or iodine-based disinfectants and cleaners can impart a medicinal taste to the treated area. Adding some wild yeast may improve the flavor and make it more therapeutic. Chlorinated water should not be used in the brewing process or to clean any equipment that will come into touch with beer. At all costs, these activities should be avoided. Water that has been treated with chlorine can be removed by using a water filter or by boiling the water for fifteen minutes and allowing it to cool at room temperature (Nishinari & Fang, 2018). Always follow the manufacturer's instructions for disinfectant usage. When used appropriately, most hand sanitizers do not leave a lingering scent. A disinfectant-treated rinse of water should be used to clean and disinfect the equipment after it has been soaked for 10 minutes in a solution containing 1/2 ounce of bleach per gallon of water. All the flavors and aromas associated with apple cider and wine are combined in one bottle of cider.

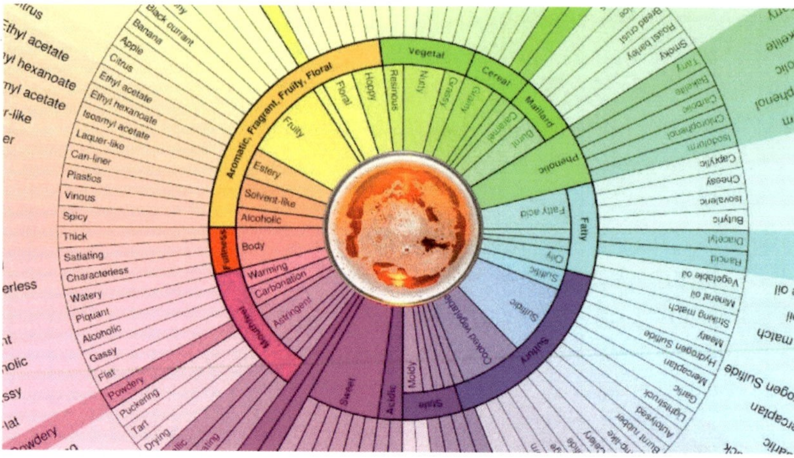

Figure 8.2: How to Identify Off-flavors in Beer.

Source: https://daily.sevenfifty.com/app/uploads/2019/06/SFD_Beer_Off_Flavors_1_2520x1420.jpg.

Incorrect usage of maize or cane sugar is the most prevalent cause of a corny or sweet aftertaste in wine or cider. For a batch of 5 gallons of cider, the maximum sugar addition is 1 pound before the manifestation of the cidery qualities. Acetaldehyde has a taste that is similar to apple cider. The sugar cane or cornmeal in your recipe may be reduced to avoid this. Using a different fermentable sugar source can reduce a beverage's cidery or vinous properties. There is no taste of cider in malt extract, whether it is dried or liquid. Another choice is honey, which may be fermented almost completely, although the intensity of the honey's fragrance and flavor will vary depending on how much is used. A laagering process may be able to help lower the strength of cidery tastes if yeast is to a fault, rather than cane or maize sugar.

The aroma and smell of diacetyl are reminiscent of rancid or over-processed dairy products, like buttered bread or brioche. In addition, the taste and texture of it are both slick and a bit abrasive.

Some of the likely causes are as follows: All yeasts naturally create diacetyl during the fermentation process, which is subsequently "reabsorbed" by the yeast cells. This may be produced by several things, including yeast with a high flocculating rate, weak or mutant yeast, excessive or inadequate oxygenation, low fermentation temperatures, weak or short boiling, and excessive or insufficient diacetyl resorption (Nwakalor & Obi, 2014). When this occurs with lagers, it is frequently viewed as a defect in the product. Some brewers and beer lovers prefer ales with low alcohol content.

What You Can Do About It: Yeast can more easily reabsorb diacetyl from wort if any of the following actions are taken: Diacetyl may be absorbed by yeast with a modest flocculation degree, though. Yeast that is extremely flocculant may precipitate out of the solution before it can absorb the diacetyl, but yeast with a medium amount of flocculation should be able to do so. If you're using a poor or mutated strain of yeast, it may not be able to break down diacetyl properly. Utilize a yeast starter and let the yeast begin its first growth. After the yeast has been pitched, it is critical to feed the yeast with sufficient oxygen for growth, but not excessive amounts of oxygen. It's critical to give the yeast enough time to finish fermenting at the ideal temperature.

Additionally, dimethyl sulfide's flavor and qualities are comparable to those of cooked vegetables, such as creamed corn, cabbage, and tomato. To convert S-methyl methionine into DMS during the malting process, heat is required. Darker base grains create less DMS than lighter ones because kilning converts SMM to DMS and removes it before it enters the wort DMS is more abundant in pale ales and lagers as a result of this. Boiling the wort, the first step in the preventive technique will drive out the DMS. A strong rolling boil must be maintained for a minimum of one hour. If you're going for maximal DMS removal, you may want to boil it for over an hour. You should never completely cover your kettle when it is boiling. This will prevent the wort from absorbing any further moisture. In addition, lengthy cooling periods might result in an excessive quantity of DMS, which can be dangerous.

To quickly get your wort's temperature down to pitching range, consider using a wort chiller or an ice bath. It is further helped by the formation of carbon dioxide during the fermentation process, which helps remove DMS (CO_2). For this reason, high-quality yeast should be used, like bubbles in gas eliminate the substance. Estery or fruity fragrance and perfume have been used to describe the aromas and flavors of bananas as well as pears, strawberries, raspberries, and grapefruits.

Fermentation naturally produces esters, which may be the root of the problem. For example, Belgian ales and Hefeweizens are recognized for their fruity aromas, and certain yeast strains generate more esters than others. Inappropriate pitching or high fermentation temperatures might cause the beer to taste or smell like fruit, both of which are undesirable for the type of beer. As the fermentation temperature rises, so does the number of esters that yeast produces. Low quantities of oxygen in the fermentation process, when paired with high fermentation temperatures, can aid in the synthesis of esters. To stay on top of it, here are some tips: Make sure you use enough yeast to attain the proper gravity and oxygenation levels in your brews. Temperatures below 75 degrees Fahrenheit are ideal for fermenting, so use them whenever feasible. Temperatures in the 60–65 degree Fahrenheit range are ideal for fermentation, although the process will take longer and produce less ester if you do (Abdulmajid et al., 2014). Make sure that the yeast you are using is appropriate for the type of beer you are brewing. If you don't want your beer to taste like bananas, stay away from yeast strains commonly used in Belgian and German wheat beers, which are designed to impart fruity characteristics.

8.3. FRESHLY MOWED GRASS WITH A STRONG AROMA

It has a grassy flavor and aroma. Some of the likely causes are as follows: It is not uncommon for grains or extracts that have been exposed to mold and bacteria to emit musty or grassy flavors and aromas when utilized. Aldehydes can build up on aging malt, and this might result in a flavor that harkens back to the grass. Hops can impart off-flavors comparable to those caused by improper handling before packing and storage. Keeping your distance from it is simple. Keep grains and extract in a cold, dry, and dark place at all times. The components should be checked for any discoloration, unusual odors, or tastes before commencing the brewing process (Muhimbula et al., 2011). When brewing using fresh grain, make sure to grind it just before using it. This will keep the grain's quality high. The grain should be used within two to four weeks of milling if it has already been done. You should only utilize the finest hops in your creations. If you're using homegrown hops, you'll want to cure them thoroughly before storing them for a lengthy period. In general, only incorporate ingredients in your cuisine that are pleasing to the eye, nose, and palate.

Tastes/smells an astringent flavor is caused by the presence of tannins or oxidation.

Figure 8.3: Different Flavors.

Source: https://c8.alamy.com/comp/G1N2KB/sign-showing-different-flavors-of-ice-cream-outside-a-shop-G1N2KB.jpg.

8.4. HUSKY/GRAINY

Over processed grains, for example, may be the source of the gritty, gruff flavor. Additionally, roasted malts can provide a smokey and grittier flavor. All-grain beers are more likely to have these characteristics than other types of beers because of the higher grain content and the additional mash and sparge time required. To stay on top of it, here are some tips: Astringency, like other tastes with a gritty or abrasive texture, needs to be handled with caution. Keep grains intact by avoiding crushing or shredding them. Using artisan roasted grains, let one to two weeks for the tastes and odors to properly emerge and be released. Foamy or granular beers that have been cold-conditioned are more prone to developing off-tastes. Cough syrup, mouthwash, Band-Aids™, cigarette smoke, and cloves are just a few examples of smells and tastes.

Possible reasons: Phenolic compounds are responsible for much of the beer's medicinal taste. Plastic, medicinal, astringent, and solvent are just a few of the taste types that phenols may create. Phosphors that taste like medicine can be created during the mashing and/or sparging process if the pH, water, and temperature are not appropriately managed. Incorrect usage of chlorine or iodine-based sanitizers can lead to the formation of chlorophenols. Ales like Hefeweizen and wheat beers have been known to incorporate cloves for flavoring purposes. To prevent having to use a sanitizer that is not suitable, you should always adhere to the sanitizer's established requirements (Munasinghe et al., 2013). If you take the same techniques to lower the number of chlorophenols and astringency, you should also experience a reduction in the possibility of experiencing medicinal tastes. When creating beer, it is important to utilize a yeast that is suited for that type of beer. In addition to having the taste and smell of coins or blood, metallic can be detected in the front and back of the mouth and throat, depending on how much is consumed.

Some of the likely causes are as follows:

- It is standard practice to boil raw metals, including iron, aluminum, and steel, to get a metallic flavor in the wort.
- A metallic flavor in a beverage can be caused by metal brewing equipment, bottle tops, and kegs.
- Using water with a high iron content will result in tastes resembling those of iron.
- If grains are not properly preserved, they may potentially develop a metallic taste.
- Brew with stainless steel equipment and containers wherever possible.
- Anything that will be used to make beer or wort should not come into touch with iron.
- Before using a ceramic-coated steel pot, check for cracks or scratches.
- There's no metallic taste to stainless steel.
- It is common for aluminum containers to lend a metallic taste to water that has an acidic pH of more than 9.

- A six-hour "bake" at a temperature of 250°F will increase the number of protective oxides that may be formed from aluminum.
- Grain that has been properly kept and is in excellent condition should always be used.
- If at all possible, stay away from iron-containing well water.
- Moldy tastes and smells like bread that has been contaminated by mold.
- A damp, humid atmosphere is almost always the cause of mold growth in beer and wort.
- When using moldy extract or grain, moldy tastes can also be added.
- Your fermenter should be kept dry and dark to avoid this.
- Your fermenter should be kept out of damp conditions.
- Ensure that all ingredients are free of smells, tastes and discoloration before beginning the brewing process (Mouta et al., 2016).
- Throw away any decayed grain.
- When malt extract is found to be contaminated with mold, it should be thrown away.
- Using the moldy extract after the mold has been removed will result in an off-tasting final product.
- Using a scraper, remove as much mold as possible from the fermenter or beer.
- Mold, on the other hand, tends to infect the entire fermenter before it can be detected.
- Oxidation imparts a distinct flavor and sent to a product.
- For starters, there's the staleness or oldness of damp cardboard; sherry; papery; pineapple; decaying vegetables; bitterness; and acridity.

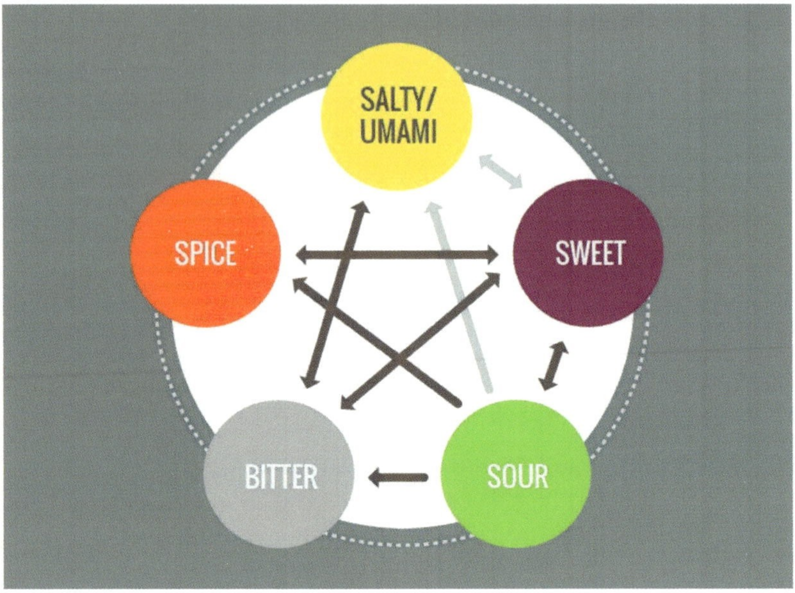

Figure 8.4: Flavor matrix.

Source: https://www.cooksmarts.com/articles/study-flavor-profiles/.

Some of the likely causes are as follows:

- Oxidation occurs when oxygen reacts negatively with the molecules in wort or beer.
- Sherry-like cardboard scents can be caused by an overabundance of oxygen in the beer, especially when the wort is still warm or after fermentation.
- Bottles with an excessive amount of headspace are more likely to oxidize.
- Aeration is necessary before the yeast can be added for optimal development and fermentation.
- To stay on top of it, here are some tips:
- An unnecessary sprinkling of fermenting beer causes oxidation almost always.

- Instead of pouring straight into the container, use tubing to decrease the amount of beer that splashes.
- To avoid air pockets, keep the transfer tube's other end below the liquid's surface.
- Take care to keep the wort out of direct sunlight (Lomarat et al., 2019).
- "Hot side aeration" refers to the practice of oxygenating wort while it is still heated.
- Because warm liquids are more prone to absorb oxygen, splashing should be avoided when the wort temperature is over 80 degrees Fahrenheit
- During and after the boiling process, splashing is not an issue since oxygen cannot be dissolved in such a hot liquid.
- Aerate the wort only when it has cooled to below 80 degrees Fahrenheit, and do it quickly.
- You should leave around 12 inches of headroom in your bottles when you're filling them.
- Oxygen absorption caps may be used to keep oxygen out of a container.
- CO_2 should be used to remove oxygen from the headspace before kegging.
- The flavor and scent of salt may be detected on the tongue's front margins while eating or drinking anything salty (Lee, 2013).
- Gypsum or Epsom salt may be used in excess, resulting in an excessively salted beer.
- Brewing salts should not be added to water until the original salt level, the effect of the salts on the water, and the right dose have been determined.
- Some beers, such as Burtonon-Trent beer, have a moderate saltiness that many people enjoy.

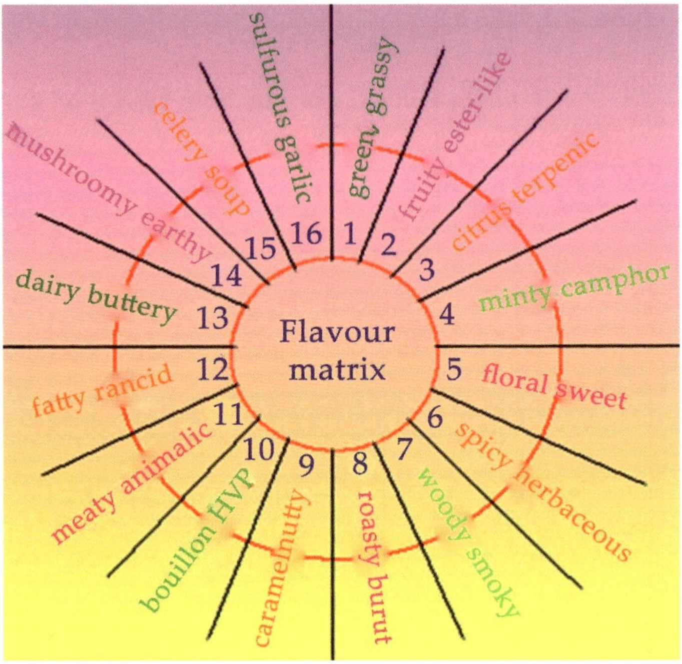

Figure 8.5: Flavor wheel.

Source: http://aris.gusc.lv/ChemFiles/ComplexCompounds/WestIndiMona/lectures/flavmap.html.

- Only use the fermenter when it is necessary.
- Use a high-quality yeast starter or a yeast starter.
- Any undesirables can be replaced or outperformed if the yeast begins to ferment sooner rather than later.
- Cleanliness is one of the most important considerations when creating outstanding beer at home.
- All Things Pleasant in Smell and Taste
- Sweet wort, syrupy, and cloyingly sweet are all examples of this adjective.

Some of the likely causes are as follows:

- When a beer tastes like an unfermented brew, it's most likely because too much yeast was used in the brewing process (Lee et al., 2021).

- When yeast continues to ferment for a prolonged period before halting, fermentation gets stalled.
- Due to the high gravity of the wort, the final gravity is much higher than it should be.
- Relatively poor tolerance for alcohol in the yeast might lead to an overabundance of residual sweetness in a high gravity brew.
- Yeast may go dormant and stop fermenting if the temperature rapidly drops.
- Inconsistent sweetness can be seen in a beer with little hop bitterness.
- Uneven sweetness is conveyed by using this term.
- Adding too many fruit flavors or other additives to a beer may also make it overly sweet.
- Make sure you use good yeast and pitch the proper amount for the gravity of wort or use a yeast starter to avoid this (Lotfy et al., 2021).
- When making beer, be sure you're using the proper kind of yeast.
- Adding enough yeast usually prevents the flocculant yeast from escaping suspension before fermentation is complete.
- To make a drier, less sweet beer, use yeast with a high attenuation percentage.
- When producing high-alcohol beers, yeast nutrients are essential.
- Keep a close eye on the fermentation temperature and don't let it go below the allowed limits.
- Gently swirl the fermenter and gradually raise the temperature to activate latent yeast.
- The option is to add more yeast.
- Even while you might stress a sweet or a bitter flavor in your recipes, remember that a beer's flavor is defined by the harmony of its many components (López-López et al., 2019).
- Add a tiny bit of fruit essence or flavoring at a time until the appropriate sweetness level is achieved.
- Pungent or sulfurous aromas might accompany bread-like tastes with a yeasty flavor.

- If the yeast used to brew the beer is ill or mutated, it might provide a "yeasty" flavor.
- An excessive amount of dead yeast in the beer can produce astringent or sulfurous tastes when the dead yeast "devours" the beer.
- As a result, immature beers may taste like they were brewed with yeast that hasn't had a chance to coagulate properly.
- While the food is being served, you may also add a yeasty flavor by pouring the yeast directly from the bottle.

8.5. AN INTRODUCTION TO HYGIENE

Many people feel that cleanliness is the most crucial component when it comes to making high-quality beer at home. If a painstakingly designed and perfectly produced beer is polluted by wild yeast or bacteria, it can develop into an unpleasant mess. Bacteria and wild yeast are likely to be responsible for nearly all of the offensive scents documented thus far (Kweku Amagloh et al., 2013). You may have an infection if none of the troubleshooting procedures indicated above work to fix the sour flavor that has formed in your brew. Make sure to sanitize all of the equipment that will come into touch with wort or beer, and follow all of the essential sanitation procedures. There is little that can be done to preserve a tainted batch, but there are lessons to be learnt. Your sanitation processes should be reviewed if your beer is tainted.

8.6. FERMENTATION'S UNDESIRABLE FLAVORS AND THEIR MANAGEMENT

Many of beer's issues stem from the fermentation process, which produces unpalatable tastes. As a result, it can be difficult to control these disagreeable scents since there are so many variables that affect how much and what type of fermentation waste is produced. In the brewing process, for example, if the wrong kinds or amounts of malt and hops are used, the beer may not have the proper taste and fragrance, but it may still be nice to drink. Even the most meticulously designed recipe might result in an unpleasant beverage if the fermentation process fails (Koli et al., 2022).

In the following paragraphs, we'll describe the five most common flavor components that are produced through fermentation. Sulfur-containing compounds such as fuel alcohols and esters are among these components.

An explanation of how these chemicals are affected by various fermentation conditions is also offered. There was no change in descriptions of wines that were clean or faulty as a result of the information that was supplied, according to the statistics. With both positive and negative words, the organic label boosted positive descriptors and decreased negative ones in wines. In addition to fruity undertones, the organic red wines ST2012 and ST2014 were noted as having animalistic or resembling undergrowth qualities (Kukovics & Németh, 2014). "Fruity" and "woody/spicey" are all qualities of a clean-tasting wine, and they characterize CT2015, organic wine from California. As a result, the organic label has been updated to incorporate new adjectives such as "fresh," "fruity," and "floral." In addition, there was a pattern in the description of animals and undergrowth, with citation percentages ranging from 6.3% to 20.8%. The original Pinot Noir was characterized by scents and tastes of "fruity" and "animal/undergrowth." When it comes to pricing, the higher "Fresh" grade assigned to Pinot Noir is the main factor. This wine's "Floral" citation increased significantly from 0% to 16.7% during the organic session.

According to CT2015 statistics, the percentage of citations for "Animal/undergrowth" increased from 58.3 to 70.8% (Abamecha, 2020). Similar to the white wines, favorable descriptions increased when flavor profiles of the organic session varied. During the organic session, the word "Animal/Undergrowth" was found more frequently in red wines than in prior sessions. The "guessing" round that took place. There was a "guess" round in which judges were given eight different wines and asked to make educated guesses about how they were made. Organic wines CB2016 and CT2015 were found by 93.8% and 87.5% of the tasters to be primarily conventional in their viewpoints, respectively. The two conventional Burgundy wines used as controls were found by 77.1% and 83.3% of the judges to contain significant amounts of organic components. The labels on the four remaining organic wines were accurate. The descriptors were combined in such a way that they matched the decision made to determine which sensory characteristics influenced the decision about the production process. Three of the twelve descriptors, as well as the emotional reaction, were found to be unique from one another at a significance level of 0.05 (Abebe et al., 2006). 51% of tasters used the term "Animal/Undergrowth" while describing an organic wine, but none of them used this term when describing a conventional wine. A higher percentage of those who believed the wine was conventional than organic used the word "oxidized" in their descriptions of the wine's aging process. 12.9% of tasters thought the wine was organic, while none of the traditional wine tasters observed the word "reductive" about it. This wine had a larger percentage of tasters who considered it to be organic than those who thought it to be conventional. When asked to estimate, tasters only recognized "Reductive" in the organic Chablis wines; but, when asked to detect it, they did so less in the organic sessions than in the conventional sessions. Fruity and "fresh" wine adjectives were more commonly employed when wines were considered conventional (Kohyama, 2020).

8.7. THE MOST COMMONLY OBSERVED EMOTIONS

In the sensory-free description, the word "unpleasant" was used, which came as a surprise given that it is not a sensory descriptor but rather an overall appraisal of emotional character. Almost all of the estimates were based on emotional reactions, except for VB2015, which had a taste that was oxidized.

In large part, this was a response to wines labeled "organic." The organic wines CB2016 and CT2015, which have a pure taste, had the least use of the phrase "Unpleasant." The answers from the guessing session might be considered the mental model of organic wine as it appears to tasters who have had substantial training and experience (Anton & Luciano, 2007). Using a word cloud made up of white and red wine drinkers' taste descriptions and emotional responses, it is easy to see what organic wines will taste like to this group. Novice tasters either favored organic wines or were willing to pay less for them when establishing the average score. Both of these patterns were spotted by observers. Free descriptions that revealed a wide range of olfactory attributes and either favorable or negative emotional reactions, such as "wonderful," "terrible," and "weird," and often compared the wines to one another were created since there was no prior education in the field.

8.8. DISCUSSION

There was a significant increase in sensory preference and a larger readiness to pay for organic products when trained tasters were educated about organic farming by organic producers (Kang et al., 2009). In the words of Apaolaza, these data demonstrate the "halo effect" of organic wine production. This is a one-of-a-kind study since the tasters were asked to evaluate ruined organic wines to see how they responded to the tainted samples. Figures reveal that even if off-tastes were discovered, consumers preferred organic wines to conventional wines by a wide margin.

Figure 8.6: There's no flavor that's off limits.

Source: https://www.biggerbolderbaking.com/best-popsicles/.

Because familiarity does not appear to be the main force behind this behavior, it is reasonable to assume that this rise in preference is the result of higher expectations. According to Ristic, freshness is related to a pleasant sensory quality, whereas a decline is linked to an odor that is less than desirable. Tasters who participated in the current study revised their descriptions of conventional wines that were meant to be "organic," such as "fresh" or "fruity," and reduced the citation of off-flavors, such as those linked with reduction, as a result of this. To put it another way, tasters altered the flavor description of conventional wines that were meant to be made from just organic grapes or perfectly organic wine. It was found by Apaolaza's research that, when wine tasters were provided knowledge about the wine's manufacturing process, their judgments of the wine's good aspects improved.

When poisoned organic wines were evaluated, there was no noticeable change in taste profiles between organic and conventional wines, which was the most notable discovery. Neither the positive nor the negative adjectives changed much throughout the tasting of the "organic" versions of these wines (Jain et al., 2014). Using terms like "Animal/Undergrowth" to describe organic wines as a whole makes sense. When compared to conventional wines with the same sensory experience, consumers preferred and were more inclined to buy organic wines. When the blind tasting was done, the organic bias may have contributed to a larger acceptance of low-quality wines. The "organic" session was also chosen over the "conventional" session because of the usage of Burgundy's traditional European wines as models for high-profile wines that are often rejected because of their first off-flavors. It was decided to utilize these wines as a model for high-profile wines that are often loathed because of their first off-flavors.

There was less use of the phrase "Reductive" for the Chablis wine during the "organic" session, but it was still used to indicate the effectiveness of that session (Igarashi et al., 2010). We assume that this contradictory behavior is due to an emotional link that makes it difficult to define how something tastes. When there was no need to speculate, more pleasant adjectives were used; nevertheless when there was a need to speculate, the suggestion of an unpleasant flavor was crucial to deciding the conclusion.

Figure 8.7: Bottles of Smirnoff vodka on the shelf in an off license.

Source: https://www.mirror.co.uk/news/uk-news/smirnoff-vodka-uk-russian-roots-26386756.

The importance of off-flavors in the recognition of organic wine was demonstrated in a tasting session in which tasters were charged with detecting the wine's production method. If a wine had no undesirable aromas or flavors, it was more likely to be categorized as "conventional" by tasters. The off-flavors in the wines and the emphasis placed on their presence during the guessing session were both signs indicating the wines were produced using organic agricultural methods. While the production procedure wasn't foreseen, off-flavors were, precisely as experts would expect in such a setting (Huang et al., 2022).

It was surprising to learn that Yeshurun and Sobel considered pleasantness an important part of the sense of smell. The term "emotional sense" was invented to describe this occurrence. Since its significant emotional weight counteracts unhappiness caused by obstacles, the organic label was more effective than fault identification. Organic certification is more useful than defect detection because of this. A relationship between liking and unpleasant sensations was not considered by Spinelli when he developed the link between sensory experiences, like, and feelings.

Since the psychological relationship to organic wines was so strong, even a defective wine detected correctly might be valued higher and earn a higher price, it was incredible. An increase in sales came about as a response to customers' belief that organic wines are generally regarded as better for them. Experts and non-experts alike have both top-down and bottom-up taste processing systems. Most likely, this increase in pleasantness is due to the same brain processes that are activated by other types of marketing, such as advertising.

Although they were told about the production process, non-experts who used a bottom-up approach did not show the same preference for bad wines. Even though they had access to the information, this was the situation. Since non-experts are less sensitive to "organic" extrinsic signals, their reactions are closer to the wine's intrinsic quality than are experienced oenology students, whose reactions are more in line with the wine's intrinsic quality (Hong et al., 2008). Only a small number of wines were included in this study, but they were selected because they best represented the off-flavors found in organic Portuguese wines. In the future, more studies should be done on off-flavors and wines from other nations.

8.9 SUMMARY OF OFF-FLAVORS

Generally speaking, seasoned tasters have a positive view of the term "organic." Before tasting, people tend to provide higher liking scores and willingness to pay ratings, which is known as the "halo effect" in sensory analysis. The uniqueness of this experiment is measuring the tolerance of tasters for off-flavors after they have been informed that the wine in issue is organic. There has been a significant rise in the popularity of organic wine because of the off-flavors that characterize it (Happiness et al., 2011). A group of oenology students came to this conclusion after correctly identifying off-flavors in wines but refusing to penalize them just because of their organic label. Wine research pioneers like Maynard Amerine at the University of California, who advocated the purity of fragrance, have a problem with this approach since it conflicts with their goal of educating students about sensory analysis. Either Émyle Peynaud or Jean Ribéreau-Gayon, are two of Bordeaux's most notable nineteenth-century academics.

CHAPTER 9

WHO CARRIES OUT SENSORY TESTING?

CONTENTS

9.1. What are the Many Types of Sensory Testing? 204

9.2. What are the Sensory Profile Requirements for the Product? 206

9.3. How are Sensorial Evaluation Samples Created? 208

9.4. Featured Facilities 210

9.5. What are the Advantages of Employing a Methodical Approach? 211

9.6. Sensory Evaluations By a Trained Panel Vs. Sensory Evaluations By a Consumer 214

9.7. The Panelists' Preference 216

The Tenamus group conducts sensory testing in its facilities using a sensory panel. According to din iso 5492, it is a group of either rookie or experienced testers. Iso 8586 differentiates between test participants and sensor subjects. Regardless of their degree of skill, all participants in a sensory exam are considered test subjects. People who are more sensitive than others are the consequence of a mix of education and life experience. A panel of tasters assessed the product (Gunaratne et al., 2019). Participants are chosen based on their abilities to recognize tastes and textures, as well as their knowledge of the grading system. Before being handed to panel members for testing in specialized facilities, meat samples are prepped and cooked in a controlled setting.

Figure 9.1: Sensory testing.

Source: https://media.istockphoto.com/photos/biologist-at-work-picture-id530484142?k=20&m=530484142&s=170667a&w=0&h=LA3jAw3RhCR4cjrOP3SY9KSlH7XTjJuXiB0f51GUF24=.

9.1. WHAT ARE THE MANY TYPES OF SENSORY TESTING?

Based on the information they offer; sensory tests are classified into three types. The three categories are discriminatory, descriptive, and emotional. The objectives of the project should lead to the selection of the right test. During a three-week rigorous training session for sensory panelists, samples, profiling, scaling, and reference are all studied. Taste facts are distinct from perspectives, which are based on personal experience, and subjectively evaluate food features based on one's personal experience (qualitative judgment). The sensory analysis uses tasting panels to evaluate goods based on how they are perceived by the human mind's senses. Tasters and items are reviewed and chosen depending on the data acquired throughout the testing procedure (Fiorentini et al., 2020). Sensory analysis and evaluation must be included in food quality programs to not only discover the primary attributes of your product but also to assure its quality and perform acceptance testing on the final product. It is vital to choose and train panelists for successful sensory panels to obtain the data required to evaluate your product and make key judgments. In this one-day interactive session, program participants learn about testing procedures and how to identify sensory qualities for new product development and quality assurance. In the food and beverage sectors, sensory testing is an integral aspect of product development. This approach may detect off-flavors and other sensory abnormalities, which aids in the analysis of product sensory profiles. A multitude of sensory tests are possible, but all rely on human panels for reliable findings (Feng et al., 2019). Smell, taste, sight, touch, and hearing are all very subjective, so what one person thinks is a strong flavor may be considered mild by another. In the face of these problems, who undertakes sensory testing to acquire objective.

Figure 9.2: Sensory testing.

Source: https://media.istockphoto.com/photos/food-quality-control-of-sea-fish-sensory-evaluation-of-sea-fish-in-a-picture-id1317558058?k=20&m=1317558058&s=612x612&w=0&h=pJWlqnqyuVZgNnMP8Mzdj-ihOuXOTSQsUy-OxHGOub8Q=.

9.2. WHAT ARE THE SENSORY PROFILE REQUIREMENTS FOR THE PRODUCT?

Who is engaged in the sensory profiling of a product?

A collection of professionals who employ their senses deliver the solution. These individuals are capable of identifying unique sensory aspects in things and providing an objective appraisal.

Sensory panelists typically have some experience in the food and beverage business, even if they lack formal training. Sensory panelists must be able to recognize and accurately describe product attributes (Elsamani et al., 2014).

A variety of sensory panels are available and are frequently used in sensory testing. Panelists with professional expertise in the sensory evaluation may be found working for sensory evaluation firms. This group has an above-average sense of touch and smell but no rigorous sensory training. Consumer sensory panelists and unskilled sensory panelists frequently collaborate on a product. Each form of sensory panelist has advantages and disadvantages, and the makeup of your sensory panel will be determined by your study objectives, project design, and budget.

Examiners must meet several criteria:

- usability;
- a disinterest in the issue of the evaluation.

9.2.1. Knowledge of a Second Language

- There were no food allergies or intolerances that preclude the product from being tested.
- The power to make oneself laugh (no olfactory, gustatory disorders, etc.) (Elina et al., 2016).

Figure 9.3: Sensory testing.

Source: https://media.istockphoto.com/photos/olfaction-test-female-scientist-examining-vanilla-smell-picture-id1320929449?k=20&m=1320929449&s=612x612&w=0&h=Wl2_57_iYo8SO4vsvcVSi3yOFkNot1ev5dvDZaiScEY=.

- The sensory test group, which includes a significant number of test participants and sensory specialists, ensures the quality and impartiality of laboratories' sensory testing. Off-flavor is defined as an unpleasant taste, smell, or mouthfeel. As a result, the sensory quality of the product degrades. This might occur during storage or processing. There is the possibility of off-flavors (Beekman et al., 2022). Bacteria can contaminate food.
- By chemical changes or contamination during storage or production; or
- By the release of undesirable chemicals

Off-flavoring can be caused by a variety of substances, and levels as low as ppb or even ppt can taint an otherwise pleasing flavor and create an analytical challenge. Consumers may be less inclined to purchase items in the future as a result of taste flaws. Sensory analysis laboratories can assist in determining why the product tastes terrible. Recommendations can then be implemented to guarantee that clients continue to receive high-quality service.

9.3. HOW ARE SENSORIAL EVALUATION SAMPLES CREATED?

Sensory testing relies on sample selection, preparation, neutralization, coding, and presentation to provide objective feedback.

- Selecting a case study.
- To limit sample variance, as many identical samples as feasible should be supplied to all testers.
- The sample's preparation.
- All samples are tested at room temperature to guarantee consistency (except, e.g., ice cream).
- Taking the sample out of the sample (Biró et al., 2020).
- There are occasions when not knowing about, say, a manufacturer is advantageous. This demands the transfer of the sample to a fresh container.
- Coding illustration.
- To protect anonymity, samples are neutrally coded before testing.
- A presentation demonstration.

Figure 9.4: Sensory evaluation.

Source: https://media.istockphoto.com/photos/close-up-professional-asian-chinese-barista-connoisseur-performing-picture-id1343280313?k=20&m=1343280313&s=612x612&w=0&h=p4niJviLCLRTzV3l8aB6Fj8uGLUjPlEx7L-YjgdGIZM=.

To guarantee a similar look and feel, the samples are placed in containers that are equal in size, shape, and color. Are there any sensory testing possibilities in our labs? In our facilities, both uncooked and cooked meals are sensory tested (Ebuehi & Oyewole, 2007). The raw materials are taste and scent assessed. Finally, our staff follows the label's instructions to recreate and re-evaluate the sensory qualities of the items.

- Dlg-based sensory assessment

The dlg scheme conducts descriptive sensory research for a product's visual, haptic (texture), olfactory (odor), and gustatory (taste) qualities. To obtain a total score, our testers record all undesirable feature features (sensory descriptions). Points are awarded on a scale of 0 to 5, with 0 representing bad performance and 5 representing great performance. Based on the results of this test, a product can be given gold, silver, or bronze classification. For a short time, the producer may use the dlg quality seal to promote his products.

- Sensory hedonic assessment

Furthermore, hedonic sensory testing is classified as "food sensory analysis." the senses may be used to assess the quality of a meal. They are distinct from professional services in that they are performed by non-professionals or repeat customers. Subjective criteria are utilized to categorize and analyze the food products under consideration using a hedonic scale. Tentamus set labs to go above and above to include a wide range of test participants to provide clients with representative data. several senses can be used to identify and analyze product qualities (eyes, nose, mouth, skin, and ears). Sensory testing is used to ensure that consumers receive a high-quality, attractive product that has been properly examined by scientists (Difonzo et al., 2019).

As a result, sensory testing is widely used in product development, quality assurance, and scientific research. They may also be used to identify if a product has reached the end of its shelf life and, as a result, if food has gone bad. Sensory testing is used to standardize and modify products to satisfy the demands and expectations of customers. Thus, new things may be offered with more success, while current ones can be improved. Sensory evaluations are done as part of quality assurance to monitor and verify the quality of a given product. The sensory assessment facility contains a food preparation kitchen/service area with pass-through compartment doors to offer food samples to taste panel members. In each booth, a touchscreen tablet is used to show questions to panelists and collect their replies, while compusense system software automates data collection and processing. On-site commercial food testing and training are also available (sensory evaluation and product development courses).

9.4. FEATURED FACILITIES

Six panelists can be seated in separate booths at the same time in the panelist area. In the food preparation/service area, a massive screen is utilized to explain the sampling approach to panelists and track each panelist's progress. In the food preparation/service room, there is a commercial range, an industrial freezer, heat lamps, and portable temperature/humidity-controlled warming units to aid in the storage and preparation of test meals (De Kock & Magano, 2020). A second food production kitchen is also available for more extensive sample preparation needs (additional information may be found at the product development facility) (refer to the product development facility for further details).

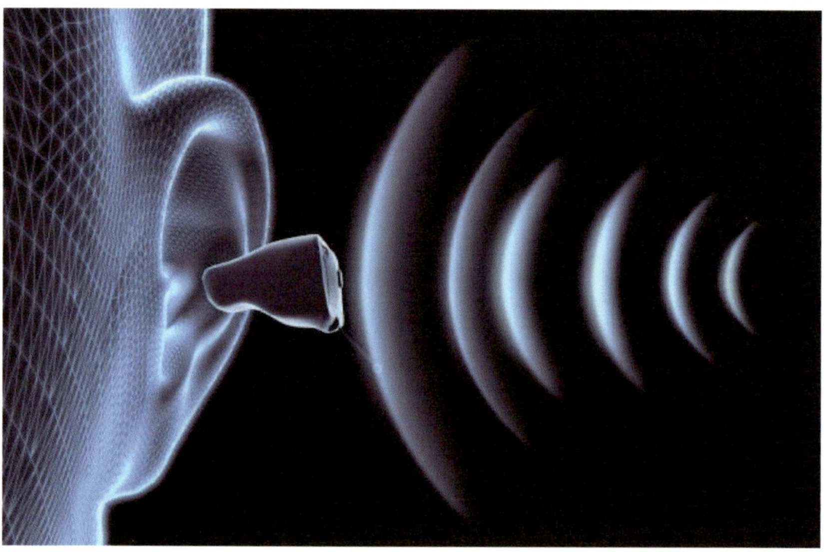

Figure 9.5: Sensory analysis.

Source: https://media.istockphoto.com/photos/ear-with-sound-waves-picture-id1253642699?k=20&m=1253642699&s=612x612&w=0&h=vV2vEpmySLmjDz4UPH3QSI3qMdj_6uTx9leBUP3qTwA=.

Color disparities between test objects can be hidden by adjusting the lighting in the panelist booths. Sensory assessment may be used to evaluate features of food and drink such as smell, taste, appearance, touch, odor, texture, and temperature. You may have been asked to offer feedback on a food or drink in an experimental setting, which might include a school or university, a workplace, or even your own home.

A team of professionals, including food chemists, taste scientists, and nutritionists, is gathered for the sensory inquiry. Consumer satisfaction and commercial objectives need sensory evaluations by food manufacturers (Daher et al., 2020).

The sensory quality of a food product may be the most crucial component for food company owners, resulting in increased product sales. When a product's sensory testing is superior to that of its rivals, its chances of success improve.

Sensory analysis may be used to translate customer demands, perceptions, and preferences into recipe and process specifications. The unbreakable laws are as follows:

Many of your clients will most likely be sharing a cheesccake at the same time. Do you believe that everyone feels the same way about it? Would they still appreciate it if they knew how low in fat it was? If their classmates rejected it due to its excessive sugar level, or if they ate it alone at home, their perspective may change (Birol et al., 2015).

According to a study, all of the symptoms we experience after consuming a portion of food have a systemic influence. As a result, each taste assessment considers factors other than flavors, such as our geographical location and socioeconomic conditions, product information such as labels, packaging, or brand, and even our cultural background and previous encounters with food items.

9.5. WHAT ARE THE ADVANTAGES OF EMPLOYING A METHODICAL APPROACH?

All of these factors influence our perception of a food's flavor, making it hard to ignore them. Our brains synthesize all of this information into a single global image of the item. With so many variables at play, how can we know what customers think about a product's flavor? In this case, a scientific foundation is essential. We employ a set of criteria and standards to restrict or regulate the effect of external components on sensory analysis (Chen & Opara, 2013).

When doing sensory testing, each object is given in a neutral setting and context. Evaluations must be done in a distraction-free atmosphere to be successful. The lighting and temperature of the space must be matched to create equal testing conditions. We serve meals on white plates, platters, and utensils for the same purpose.

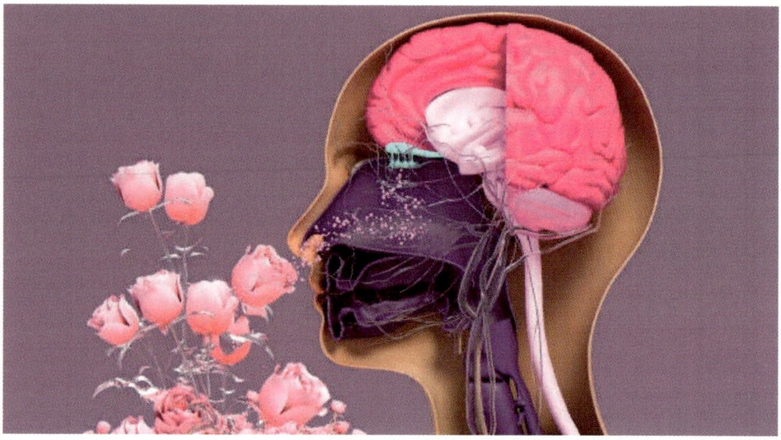

Figure 9.6: Sensory senses analysis.

Source: https://media.istockphoto.com/photos/olfactory-system-sensory-system-used-for-smeling-olfaction-senses-picture-id1302713332?k=20&m=1302713332&s=612x612&w=0&h=WLMnJjZBHN2hms4PfgsT4-PP6mioK3AONNROpI_PPcQ=.

To obtain valid findings from sensory testing, the following evaluation standards must be met:

- During assessments, sensory judges must act independently and in isolation from one another. Individual judgments will be obtained independently of the group. Individual cabins have been created in the sensory laboratory for the judges to operate in private.
- The panel of judges will never be able to figure it out. As a result, each sample is issued a three-digit or three-letter code (Chowdhury & Das, 2015). No product-related information is permitted to influence the judges' decisions in this way.
- The order in which the samples are presented may have an impact on the judges' choices. To avoid affecting the entire panel, the presentation sequence must be randomized and vary among judges.

- To avoid misunderstanding or prejudice on the side of the researchers, all questions and responses from the judges are presented in a standardized form.

This scientific approach is critical for examining the emotions associated with product consumption. We must never forget that our subjectivity and sensitivity extend well beyond the taste of a thing!

Analytical research necessitates increased sensory sensitivity. This means that all research participants are likely to be more sensitive to differences in the items under consideration than the ordinary consumer. As a result, panel members should be chosen through a series of screening activities. In these activities, discrimination tests are commonly used, and the rate of correct responses for acceptable subjects should be significantly greater than the chance probability for the tests used (either p=1/3 for triangle tests or p=12 for duo-trio tests). Many ways are used in practice to test a person's sensory sensitivity, depending on the severity of the screening exam or the needs of the project (Bodyfelt et al., 2008). Furthermore, sensory testing should not be needed for those who have a prior technical understanding of items or projects. 1 with these attributes in mind, panelists for descriptive assessments must also possess extraordinary linguistic abilities, which are required for building a common language during training sessions.

Figure 9.7: Sensory evaluation.

Source: https://media.istockphoto.com/photos/selective-focus-of-coffee-shop-workers-checking-coffee-quality-during-picture-id1303762877?k=20&m=1303762877&s=612x612&w=0&h=4KCZpsP2rSVXBzEGDArZn5-Z5h5BRhi-joZc213qVzXM=.

Affective testing seeks to differentiate between things based on consumer preferences. Affective evaluation. Everyone who participates must be actual product consumers and approvers, capable of indicating their degree of approval for various product categories in several methods. Subjects who score every item as either strongly favorable or overwhelmingly negative are less likely to aid researchers (Chira et al., 2020). These responses can be found during screening sessions and excused from real product testing. They should not be included in acceptance testing to prevent persons who are familiar with the goods or projects and have been trained in analytical testing from causing difficulties.

9.6. SENSORY EVALUATIONS BY A TRAINED PANEL VS. SENSORY EVALUATIONS BY A CONSUMER

Sensory assessments are typical practice when new items are introduced, when products are changed, when products are marketed, and so on. They're useful for more than just pre-packaged meals. Consumer panels and trained sensory panels are utilized in addition to fresh and dried fruit.

For sensory evaluation, both consumer and trained panels are routinely utilized. Examples of panels I've worked with throughout my graduate studies show the distinctions between the two. These gatherings are a lot of fun for both the organizer and the participant!

Is it feasible to explain the operation of a consumer sensory panel?

Using surveys to learn more about potential clients' interests and requirements. These exams are taken by people who have no relationship with one another. As a consequence, they may also tell you whether or not customers can distinguish between different product kinds. They are often seen at supermarkets. Switching from artificial to natural food coloring is an example of a basic experiment to see if the client perceives a difference. This dish may be made using both fresh and dried fruits (Coggins, 2007).

One of my colleagues' projects at UC Davis focused on mango ripening, where he investigated a wide range of ripening phases at varied temperatures. During her investigation, she held focus groups with potential consumers to determine whether or not: Customers can distinguish between cooking techniques depending on flavor.

Which technique did you favor when it came to selecting one?

In addition to consumer panel data, mango skin and flesh color, sweetness (SSC), and acidity (ta) were examined. After the study, all of the data acquired could be utilized to identify which mango variety was the most popular and why.

Figure 9.8: Olfaction.

Source: https://media.istockphoto.com/photos/rapid-antigen-nasal-test-could-hurt-picture-id1318837169?k=20&m=1318837169&s=612x612&w=0&h=by hSMCyBh5Fi9Xo6LqIA9aQnCrqxy1mB-m53PhqJuuQ=.

What is the purpose of a trained sensory panel?

Using trained panels to recognize and distinguish between product variations. Consumer panels aim to determine what people want, whereas trained panels do not. Each panelist has gone through the same training and is familiar with the descriptions. Following training, panelists should be able to distinguish between goods (as a group or individually), characterize the various product features (qualitatively), and assess the severity of the attribute (as a group or individually) (quantitatively) (quantitatively). The success of a taught panel is determined by the panelists' ability to detect differences, replicate their results, and be consistent with the rest of the taught panel (Chen et al., 2022). Furthermore, the testing environment differs from that of a consumer panel. Panelists must be in a controlled environment with specific lighting, temperature, and booths.

9.7. THE PANELISTS' PREFERENCE

The panelists are selected based on the type of test required. Consumer panelists, as well as professional or experienced panelists, answer a variety of sensory questions. Consumers the consumer group can be distinguished based on geography, age, socioeconomic status, or any other population characteristic. Customers that visit a certain location, such as a supermarket or shopping mall, are often chosen at random to serve as panelists. This information may be used to indicate consumer preferences, acceptability, product usage, and feedback on wine samples. In terms of marketing and sales, this might be the most important piece of information about the wine. Customer testing is both time and money-consuming, yet it is critical for analyzing customer opinion.

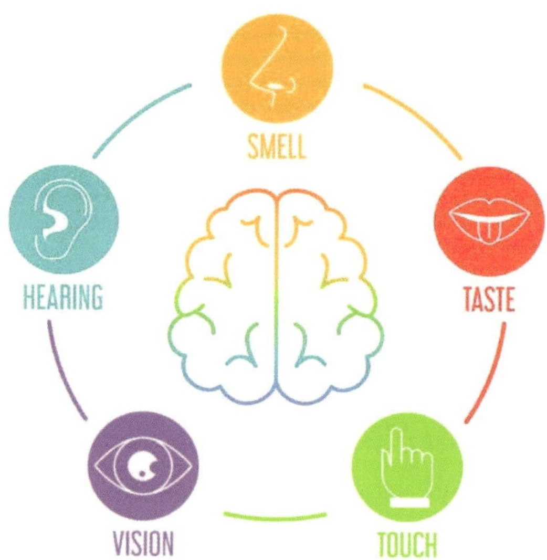

Figure 9.9: Sensory senses.

Source: https://media.istockphoto.com/vectors/five-human-senses-icon-set-vector-id1249129926?k=20&m=1249129926&s=612x612&w=0&h=lkbWSkaJ___C9fGEZR0aoUzAq1XUhqOs5z_KtHYIvjs=.

CHAPTER 10

SAMPLE PREPARATION FOR SENSORY EVALUATION

CONTENTS

10.1. Effective Sensory Evaluation 218

10.2. Understanding How Our Environment Defines Your Tasting Behavior 225

10.3. Preparation of the Specimen 235

10.4. A Category's Extensibility 238

10.5. Descriptive Tests Can Be Used to Evaluate Cookies 239

The "standard sensory practices," or protocols and practices for accurate and reliable testing, must be addressed. Sensory evaluations are frequently conducted in controlled environments to reduce panelist bias, but these facilities are not always readily available, requiring those conducting the assessments to use alternative methods. These common-sense practices can help operators choose the best sensory evaluation test for their needs and improve testing in any setting. There is no need to measure anything in the case of sensory examinations, also known as attribute examinations. In contrast to measuring tests, variable tests, numerical values are only assigned after qualitative results have been provided when performing qualitative tests, e.g., profile analyzes, dig test. Chemical-physical measurement methods such as ph. value measurements, spectrophotometric methods, and mass spectrometry are considered more reliable than sensory tests. It is widely assumed that they are ill-defined, time-consuming, and produce few results (Castro et al., 2013). Because small and medium-sized businesses have fewer personnel resources, they either eliminate or use deceptive "tastings" to promote these biases. Sensory tests are useful for market research, product development, and quality assurance because they require careful planning, preparation, execution, and evaluation. In real-world situations, errors are more likely to be the norm than the exception at any point during these phases. As a result, there is plenty of room for expansion.

10.1. EFFECTIVE SENSORY EVALUATION

- Having a specific goal in mind, when conducting a sensory test, it is critical to understand the problem and the purpose of the test: Sensory analysis can be used to measure the intensity, difference, and other properties of food using a device such as a gore or HPLC, food sensory analysis.
- Consider the public's attitude, sensory analysis, individual sensory analysis, of a difference, flaw, or other distinction is referred to as measurement. Measure of consumer acceptance or preference, hedonic, affective, sensory evaluation it must first be determined what the target is to plan a sensory test.

Preparing for the exam Sensory evaluations should consider the following:

- Sample preparation and presentation Testers
- testing conditions, testing method, and

- Testing time/process to ensure accurate and repeatable results, additional specifications must be developed (Castro et al., 2013).

Consider the following statement when designing tests: Three effective sensory tests were planned, administered, and evaluated Samples, sample selection One or more food samples could represent the random samples to be evaluated during a sensory test.

Figure 10.1: Sensory evaluation.

Source: https://media.istockphoto.com/photos/close-up-professional-asian-chinese-barista-connoisseur-performing-picture-id1343280313?k=20&m=1343280313&s=612x612&w=0&h=p4niJviLCLRTzV3l8aB6Fj8uGLUjPlEx7L-YjgdGIZM=.

Alternatively, one or more representative samples are gathered from the broader public Sensory testing is done on random samples of food from a range of persons, such as the dlg test, the profile test, or the in/out test ,batches, varieties, formulation modifications, etc. It's vital to remember that random samples should be representative of the population while selecting and drawing them. To evaluate whether or not customers would detect a change in the formulation of food already on the market, a sample of consumers would be employed. A representative sample of consumers is essential to draw generalizations about the complete consumer population;

(or of a sub-group such as children in the case of a meal principally intended for and consumed by children) Comparable to a preference or acceptability test (Yağcı & Göğüş, 2008). Food samples utilized in the hedonic test research must be as representative as possible of the various variants of the formulation under examination. Participants in the research, as a result, the test's goal determine how many testers will be assigned to it.

Based on the test criteria, certified sensory testers or sensory specialists are employed to conduct food sensory evaluations. Instead of limited familiarity with a single product or range of items, the latter has an extensive and, in most circumstances, wide understanding of the subject matter at hand. Most sensory tests are done in groups, test panels, to lessen subjective effects while boosting individual testers' ability, panel as a "broadband test instrument" It is not always possible, however, to make decisions solely on the output of the group. No matter how many times the food item is evaluated, the problem will still be there even if only a few testers identify it, e.g., an overdone or musty flavor or scent, an overcooked or musty flavor or aroma. A gas chromatography or high-performance liquid chromatography (HPLC) facility acts as the tester here, and if numerous instruments with variable measurement sensitivity are available, it is always advisable to use the most sensitive one.

More particular information on the composition and assignment of test panels used for sensory food analysis may be found in dlg expert papers on panel selection and certification. Hedonic tests should never be provided to qualified testers. Because of their prior profession as sensory testers, they are no longer considered "regular" clients. Untrained test subjects must always be exposed to acceptance and preference tests, as well as tests that analyze consumer perceptions of, for instance, product faults or inconsistencies. To obtain representative data, large samples are required due to the vast variation in sensory sensitivity, sensory preferences, and sensory aversions among consumers. As a general rule, only consumers who are familiar with and often consume the product should be invited to participate in a test to evaluate if the product's sensory characteristics have altered as a consequence of a change in the formulation or production technique (Yang & Boyle, 2016). The target audience of the product should be taken into account when selecting a customer sample when seeking to measure the interest in, or acceptance of, a new product or when attempting to pick the preferred version of numerous competing products. Then, a random sample will be picked from this group. Consumers may be randomly picked for a certain test or taken from an existing pool of prospective test subjects, such as those retained by major organizations and commercial testing institutes, depending on the exact test objective.

Conditionality's for Evaluation the purpose of a sensory examination is linked to the setting in which the test is conducted. Analytical testing should be conducted in a manner that does not negatively affect the testers' duties, does not distort test findings, and provides repeatable results. Suitable room settings "sensory analysis basic recommendations on the design of test rooms" and technical equipment, such as air conditioning, is essential to manage and maintain test circumstances. A professional sensory laboratory is an ideal solution for these sorts of applications. Because it encourages testers to focus on the product and "study" it in sensory terms, this atmosphere is not indicated for hedonic testing. Results may differ considerably from those that would be expected under usual use. As a result, the conditions of the research should be as near as practical as those of a real meal (Warnakulasooriya et al., 2020). It is feasible to do the test at a shopping mall, a consumer trade exhibition, or any other place where food samples can be delivered to the testers for home use (home use test) (central location test) However, under these circumstances, it is impossible to coordinate or implement any critical product preparation such as soup in a package.

Figure 10.2: Sensory analysis.

Source: https://media.istockphoto.com/photos/professional-q-grader-preparing-to-test-and-inspecting-the-quality-of-picture-id1210600057?k=20&m=1210600057&s=612x612&w=0&h=UeOmKiEC37sWohLnQKYcwUEBQ-RT-4DzWPZ7Ype6uyrA=.

Permanent or transportable test studios equipped with the appropriate instruments are a feasible solution. Test Methodology, the testing technique is also dictated by the objective of the sensory evaluation. To determine how many samples to test, what type, e.g., spicy or hot are available, the time allotted and how many testers are needed are just a few of the many variables that must be taken into account. As an example, the detection sensitivity of the various methods for identifying sensory differences varies, as does the necessity for samples, the number of testers necessary, the cost, and the length of the test (Tsai et al., 2021). When performing a profile analysis, panelists are expected to follow additional requirements than when undertaking a conventional descriptor test. Sample preparation and presentation all of these aspects are taken into account during testing, from sample preparation to display layout. Keeping external influences, known as context effects, from altering test findings and causing them to be erroneous will be crucial in the future.

Other sensory features that are different from the one being assessed could cause context effects, as might sample type, i.e., the subset that is picked. The container used to store and test food the number of samples delivered, the manner samples are grouped, the order in which samples are assessed, the testing set, the presence of other persons or superiors, and the testing environment. Some sensory qualities may be interpreted differently depending on a food's appearance. There is a belief that fruit juices with a darker hue are more fragrant than those that are lighter in color. You shouldn't estimate an odor, taste, or texture differentiation only on the sight of a different test. Such discrepancies can be hidden in test cabins by utilizing colored lights (red, blue, and yellow). If required, the testing room may be darkened and the testers might be rendered entirely blind (Tuorila & Monteleone, 2009). Carrots and apples could taste different if their texture changes. It is feasible to examine the juice or grated or rubbed samples in such cases. An assessment of a food's taste and aroma must involve nose clips.

The impact of inhomogeneous meals could be substantially different depending on the subgroup. A sampling method that ensures testers are always dealing with comparable data is vital. Furthermore, the samples should not be used to draw any conclusions about the manufacturer, the brand, or anything else. Hedonic tests and quality evaluations would be invalidated as a result of this. This can be done by utilizing neutral, identical test vessels or by anonymizing the original packaging, something that cannot be done with containers that are proprietary to a single company or brand. There are at least three character codes, preferably four to five, that are randomly generated numbers and letters that are used to identify the samples.

Figure 10.3: Sensory laboratory analysis.

Source: https://pan.olsztyn.pl/2017/10/training-at-sensory-laboratory/.

The sensory properties of neutral containers must not be modified if samples are transported to them. The temperature of the samples must be suitable for the cuisine as the right consumption temperature is strongly tied to the quality and popularity of the dish. At 20 degrees Celsius, beer, for example, is poorer than beer served at room temperature. Changing temperatures may generate variable perceptions of the intensity of individual sensory aspects (profile analysis) or differing judgments of sensory qualities when a large number of samples are being examined.

Whether or if the dish was regarded favorably. Temperature is used by testers to estimate the sample that deviates from the reference sample or the reference sample that matches the deviating sample while performing a difference test (Short et al., 2021). A sufficient quantity of samples must be supplied to the lab to complete the test and allow the testers to retest their results.

At the same time, each sample should be distributed evenly to the testers. Due to lack of equitable opportunities to re-taste results in a range of situations, ever tester may be prevented from having the same experience. Testers may also feel that a smaller sample indicates a greater sample cost, which may affect the quality or appeal of the tests. The sequence, in which the samples are examined, as well as the positioning of the samples on a tray, must not alter a sample set or several sample sets. Even if a person's sensory performance improves during the exam, it may drop as a result of fatigue or adaptation. The results of prior tests may impact testing to evaluate the quality or popularity of a sample. All samples must be organized in the same sequence for each tester, such as one row after the other, and evaluated in the same order, such as left to right.

There must be some type of randomization, e.g., by rolling dice or using a computer, in the sample set for each tester, or there must be some sort of balance in the sample set. In the latter scenario, consider all conceivable sample configurations. A triangle test, for example, can provide six different configurations. As a result, selecting examinations that can be separated into six sections is strongly advised. If there were 30 testers, each of the six potential layouts would be tried five times. In a paired comparison or paired preference test, half of the testers will evaluate one of the two alternatives (Ayzeli et al., 2020). To eliminate bias, testers must also be provided random samples, e.g., by drawing lots. It is hard to choose the order of inspection of a large number of samples or sets without relying upon chance or equilibrium.

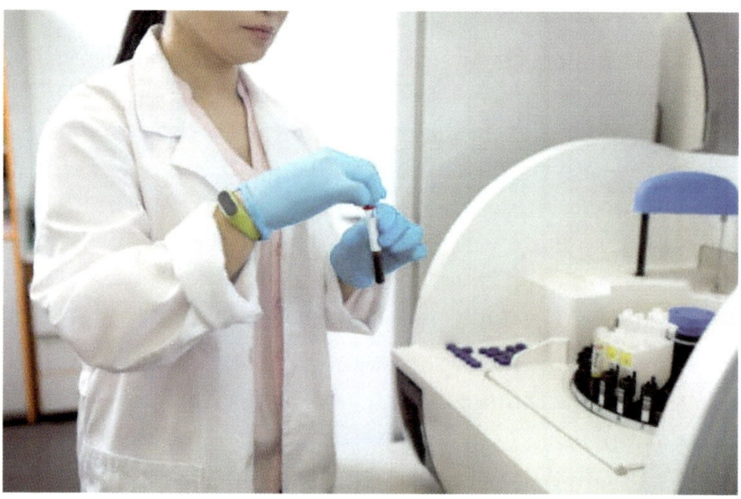

Figure 10.4: Sample preparation.

Source: https://media.istockphoto.com/photos/darkhaired-doctor-in-a-white-robe-holding-the-test-tube-picture-id1207324914?k=20&m=1207324914&s=612x612&w=0&h=Gt6SNoM9NWotExkt5ZY-bezRMn2rQp31plIFwSlAcyo=.

The samples analyzed in many sessions are identical. Because tester performance skills, test settings, and other circumstances may differ between sessions, the notion that the date and time of a session has a systematic impact on test findings must be disregarded. An adverse testing setting should not influence the test results Unfavorable circumstances include inclement weather and loud sounds. Foreign odors, poor lighting in testing rooms, and vividly colored walls or furniture may all mislead. Maintaining a constant test environment assures that the test's outcome is not altered by the settings under which it is run. The setting may also be influenced by the test administrator or another individual present during the test (Sutrisna et al., 2017). The influence of testers' comments, facial imitation, and gesture imitation may mislead testing findings. There might be an issue with the panel's hierarchy. Subordinates' decision-making abilities may be impaired if they are on the same panel as their superiors.

As a result, it is critical to avoid such alignments. Manufacturing employees' knowledge of the ingredients or, if relevant, what went wrong throughout the process might impact the context of a meal. The findings of the analyzes are also provided to the lab personnel. The panelists may assume they can taste certain components or detect flawed attitudes, but this is not the case. It also influences how the product's quality is assessed. Administrative and maintenance personnel who have never dealt with sensory analysis before should form the panel for an internal sensory analysis test. Client requests, perceptions, and preferences may be used to generate recipe and process specifications using sensory analysis. The unbreakable laws are as follows:

10.2. UNDERSTANDING HOW OUR ENVIRONMENT DEFINES YOUR TASTING BEHAVIOR

Consider how many of your customers are enjoying a cheesecake at the same moment. Is it something that everyone equally values? Would people still appreciate it if they knew it had less fat? If their classmates rejected it due to its excessive sugar level, or if they ate it alone at home, their perspective may change. According to a study, symptoms experienced after eating a meal have a systemic impact. This implies that, in addition to the flavor, our location and social surroundings; product information such as labels, packaging, or brand; and even our culture and previous experience with food items are all considered when assessing a flavor. Why is it vital to employ a scientific method? It's difficult, if not impossible, to disregard all of these impacts on our taste senses when it comes to food (Schiling & Pham, 2012). To gain a comprehensive picture of anything, the brain gathers all of the facts it can get its hands on. When there are so many unknowns, how can we know what customers think about a product's flavor? In this case, a scientific foundation is essential. We use a set of criteria and guidelines to restrict and/or regulate the effect of any extraneous components on sensory analysis. During a sensory test, the items are always shown in the same neutral setting and with the same controls in each participant's brain.

Figure 10.5: Sensory evaluation.

Source: https://media.istockphoto.com/photos/wine-tasting-event-picture-id505914328?k=20&m=505914328&s=612x612&w=0&h=KqwWcWOP-zT-FuQUt1EhsTKJc8AW1QvOORtejPV8R4P4=.

Distracting noises and odors must be avoided in the testing set. Only by standardizing both lighting and temperature in a test room can consistent testing conditions be obtained. This is the most typical reason that white serving trays, plates, and silverware are used. Specific grading standards are specified. If you want reliable sensory testing results, you must adhere to the following evaluation criteria: Each sensory judge must operate alone and without interacting with other judges when completing a sensory examination. This ensures that each individual's judgment is collected free of the influence of the group. Because they are unable to identify the things, the judges are limited to the sensory laboratory's private cabins, which are meant to encourage solitary work (Shinde et al., 2016). As a result, each sample is issued a three-digit or three-letter code. This prevents the judges' decisions from being influenced by information about the product itself. The order in which the samples are presented may impact the judges' decisions.

To control this impact on the full panel of judges, the presentation sequence must be random for each judge. To eliminate misunderstandings, misinterpretation, and researcher bias, researchers and judges adopt a uniform set of questions and replies. This meticulous scientific methodology must be used to study emotional responses to product consumption. We must never lose sight of the reality that our sensitivities are subjective, no matter how much we adore a product's flavor. Sensory assessment is rarely exploited to its full potential in the workplace when it comes to making business decisions. Sensory assessment is taught in such a way that pupils do not associate appropriate test circumstances with industrial concerns or challenges. To solve the issue, a step-by-step approach for conducting sensory evaluations was developed. These are the six stages of performing sensory evaluations: The initial step is to look into the sensory problem or issue the next step is to determine the exam's aim. This is the final and final step. In this phase, you will put the test to use. Assess the results. You must act on the sixth and final stage. To complete a sensory evaluation of a problem or inquiry, six stages are necessary. Even while it does not provide answers for all possible sensory evaluation applications, it does assist the user in making the optimal choices.

A smart way to start is by answering the what, why, when, where, and who questions concerning a problem or inquiry. Using these responses, we may establish the major purpose of the exam. To perform a successful sensory test, the test's aims must be well defined. Once the test aim has been established, the most relevant test parameters, such as the test technique, the test environment, the test panel, and the test samples, may be chosen. Effective sensory testing necessitates meticulous planning. It is vital to choose an acceptable method for interpreting test findings. This is the final part of any sensory exam, and it involves analyzing the original question, the test objective, the test settings, how the test was performed, and the interpretation of the results step .Case studies will be used to demonstrate this strategy. Humans can employ their senses of sight, hearing, smell, and touch. The four senses of humans are sight, hearing, smell, and touch (Sun et al., 2021). People employ all of their senses when shopping for food: sight, smell, touch, and taste. When it comes to buying, preparing, and eating a meal, various aspects impact consumer views, including pricing, packaging, look (raw or cooked), and ease of preparation. The most essential component in forecasting future purchases is sensory quality.

Untrained sensory panels and consumer-oriented testing procedures can be used to learn about customer preferences. Product testing is required to identify the specific sensory properties of a meal. The detection and assessment of sensory attributes are required to generate novel food products or re-formulate existing commodities. This involves detecting and measuring changes caused by processing, storing, or adding new components. This form of quantitative data is collected in the laboratory by properly trained sensory panels, because, when refining food formulations or developing new recipes, product-oriented testing often takes precedence over consumer testing. When doing actual consumer testing, it is critical to choose a large random sample of people who represent the target community of potential consumers. Panelists should not be taught or chosen based on their sensory acuity.

Figure 10.6: Food sensory test.

Source: https://media.istockphoto.com/photos/food-quality-control-of-sea-fish-sensory-evaluation-of-sea-fish-in-a-picture-id1317558058?k=20&m=1317558058&s=612x612&w=0&h=pJWlqnqyuVZgNnMP8Mzdj-ihOuXOTSQsUy-OxHGOub8Q=.

Typically, 100 to 500 people are surveyed or interviewed for this form of research, and the results are used to forecast the opinions of the target population. For interviews or testing, customers' homes or a large site, such as a market, school, retail mall, or community center, may be used. When performing a real consumer test, selecting a representative sample of the target group is time-consuming and costly. As a result, untrained in-house customer panels are regularly utilized to generate preliminary product acceptance figures, frequently before the completion of true consumer testing. An in-house panel provides the organization with greater control over testing methodologies and settings. True consumer tests should be supplemented not replaced, by in-house panels (Serventi et al., 2022). Small trained panels serve as testing equipment in product testing. Trained panels are used to discriminate between identical objects or to rate the quality of flavor (smell and taste), texture, and visual attributes. These panels are normally made up of five to fifteen people, each of whom has been trained for the specific task at hand based on their level of sensory sensitivity. A trained panel should not be utilized to judge the acceptability of a meal.

Because of their particular training in disregarding personal preferences, sensory metrics are more sensitive to minor deviations than the ordinary consumer. While a substantial infrastructure is not required for sensory testing, several conditions must be provided if tests are to be done correctly and the findings believed. Even if permanent sensory testing facilities are not available, current laboratory space can be adapted to enable sensory testing. To begin, all sensory testing facilities must contain a food preparation room, a separate panel discussion area, separate quiet panel booths, a desk or office space for the panel leader, and supplies for creating and presenting samples. Various sensory laboratories with permanent sensory testing facilities and designs have been built. In the kitchen, there is a preparation space (Silva et al., 2020). In the food preparation area, worktops, sinks, cooking and refrigeration equipment, and storage space are all required. Spaces that are well-lit and ventilated are essential. There must be adequate counter space to prepare meals and store sample trays before presenting them to the panelists. The recommended counter height is around 90 cm or36 inches. The depth of a conventional counter is 60 cm or 24 inches.

Sinks; There should ideally be at least two sinks with both hot and cold water. In the sensory laboratory, distilled water is also useful. When cooking, cleaning, and washing dishes, use distilled water to remove any odors or tastes. Kitchen equipment Separate heating components and ovens must be conveniently accessible, whether powered by gas or electricity. Microwave ovens may also be useful in the kitchen. Refrigeration equipment and systems Refrigeration is required for the storage of perishable items, and samples may need to be chilled to a consistent low temperature before serving. A separate freezer might be useful for storing long-term components, reference samples, and meals prepared at different periods. There is plenty of space for storage. Cabinets or closed shelves for dish and supply storage should be installed beneath working tables and above pass-through panels.

During panel construction, trays can be stored on an open shelf above the pass-through component. Napkins, pencils, plastic spoons and forks, and other panel items may be stored in drawers beneath the countertops. Ventilations, cooking odors should not enter the panel room, thus ventilation hoods with exhaust fans should be installed above the burners. A discussion panel is provided for product testing. It is vital to provide a location where panelists and the panel leader may gather for education, training, and debate (Abdulmajid et al., 2014). By totally segregating the discussion space from the kitchen's noise and aromas, the panelists' work should be protected. It should be placed such that it does not interfere with other laboratory personnel. In a comfortable, well-lit area set up a large table and enough chairs or stools to seat at least 10 visitors. The discussion panelists should have access to a big chalkboard, flip chart, or whiteboard. The bulletin board beside the entryway might be used to post notices and information about the panelists' performances.

Figure 10.7: Sensory evaluation.

Source: https://media.istockphoto.com/photos/professional-asian-chinese-barista-grader-performing-coffee-cupping-picture-id1338803468?k=20&m=1338803468&s=612x612&w=0&h=oZNy0FcNHmkTVJ_SWw9af6m7Re3W-fR25AEOBwBa_bYs=.

This part has a panel booth. It's ideal to keep the booth and kitchen separate, much as the dining table. Although a separate panel booth room is preferred, sections can be merged by placing the booths along one side of the discussion room without a dividing wall. However, group discussions and one-on-one sample sessions cannot take place at the same time. This arrangement may be problematic if multiple sensory activities are taking place at the same time. There are several sensory evaluation exams available. A one-time event Thomas used a single sample test to evaluate goods with a distinct flavor or aftertaste, such as coffee. The evaluator receives only one sample of a food item at a time during single sample testing) (Rousta et al., 2021). Because only one food item is evaluated at a time, this is not the most efficient test. A three-way evaluation A paired comparison compares two meals. It's a paired comparison test in which two or more items are compared after one ingredient or component has been modified.

The evaluator in a duo-trio test is unaware of the identity of the samples being examined at the time. Thomas may use a 70% dark chocolate sample as a reference for this test and compare it to a second 70% dark chocolate sample and a 70% dark chocolate sample with additional salt. He'd then determine which sample was identical to the reference sample. The duo-trio tests can determine whether a product's flavor changes as a result of changes in its components or makeup. Triangles are being researched. The triangle test is another sort of paired comparison exam. Thomas would use the triangle test to see whether two things can be separated from one another. This sort of test employs three samples, two identical and one unique, to get three outcomes.

There are two pieces of dark chocolate with 70% cacao and one piece of chocolate with 60% cacao. He would be the one to determine an irregularity. The sensory testing setting will differ depending on the operator, but particular location selection criteria should be followed to avoid panelist bias, whether the venue is an indoor, climate-controlled room or the rear of a hop yard. To avoid interfering with the results, noise and odor should be kept to a minimum during testing. For example, if hops were tested in a running brew house, the fragrance of beer would be too strong. Panelists and operators should be able to quickly access the testing area; therefore a location that is not in the heart of a busy farm or processing business should be considered (Pu et al., 2020). Panelists should not have access to any potentially biased information. Panelists may give inaccurate marks to samples if they observe the production or preparation processes for the objects they are examining.

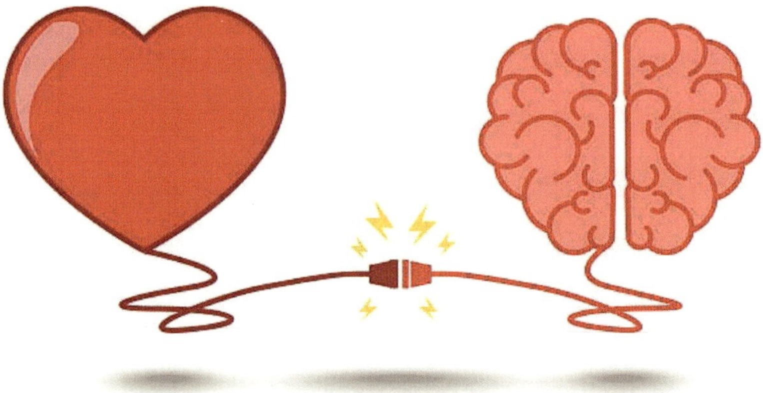

Figure 10.8: Sensory integration.

Source: https://media.istockphoto.com/vectors/brain-and-heart-interactions-concept-best-teamwork-vector-id536652254?k=20&m=536652254&s=612x612&w=0&h=oV_Lu4qRkM0wwBqc1WK87f-0zzLPh3OrVEm9W4UqLFo=.

Access to the outcomes of other panelists must be restricted as well, as this may accidentally impact their decision-making process. In formal sensory testing facilities, panels with built-in booths are commonly used to segregate panelists and distribute samples from a prep room through hinged doors. In unusual testing conditions, temporary booths made of plywood or trifold poster boards can be utilized to segregate panelists. Finally, the testing site should have sufficient preparation facilities. This might include having access to water, refrigeration, sample processing, and storage for all essential supplies and equipment, depending on the crop or product being tested. Existing panels may be able to support numerous sample sessions if the testing is large enough, while this is less of a concern for testing on a small or medium scale. Random selection/labeling in a haphazard manner as part of the blind labeling sensory evaluation technique, samples are randomly allocated a three-digit identification code.

To avoid the panelists noticing any patterns or sequences in the sampling process, samples are shown randomly rather than in any specific order. Because random number generation by hand is never completely random and always contains patterns, it is preferable to utilize online tables or random number generators. Depending on the testing procedure, the sample order among panelists should be randomized. In the first sample, an equal number of panelists should be given Product A and Product B. Testing may commence when an appropriate procedure for the crop or product has been chosen and randomized. A sample ballot should be prepared once each sample has been assigned a three-digit label. The title is as follows: The title may include the name of the testing technique, a brief description of the product, or both (Abamecha, 2020). To minimize prejudice, ensure that the title contains only enough information for panelists to determine whether or not to participate. In the test title of a commercially marketed alcoholic beverage. Because the panel will be hosted on a university campus, the alcohol percentage is an important factor for possible panelists to consider because some of the panelists may be young or have alcohol intolerance.

Depending on the purpose of the experiment, it may also represent the flavor of the product being evaluated. Finally, any brand names in the title that might raise suspicion should be removed. Panelists should be given clear and concise instructions that include all of the information they need to complete the review correctly. All test instructions should provide suggestions for sample evaluation or sample sequencing, as well as the feature being tested, if applicable.

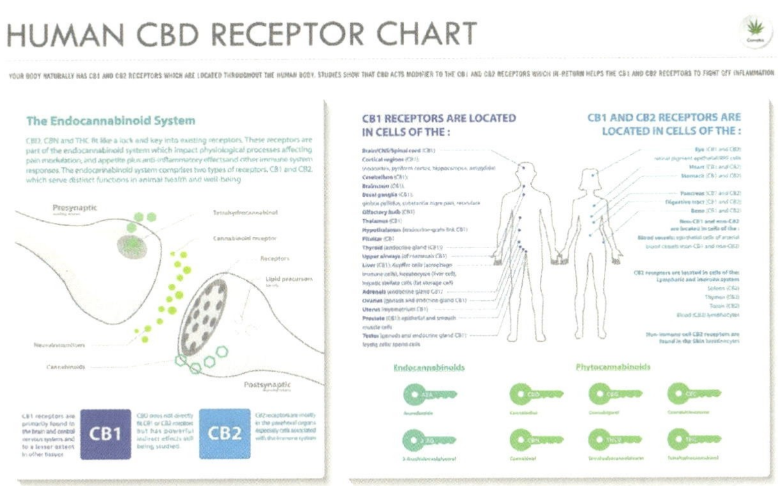

Figure 10.9: Sensory evaluation image.

Source: https://media.istockphoto.com/vectors/human-cbd-receptor-chart-horizontal-business-infographic-vector-id1176590944?k=20&m=1176590944&s=612x612&w=0&h=etnCb689RJP10TDzC0kUDPPDuaai1lAW71-03TOYK1M4=.

Some exams are easier to understand and require less explanation than others. 3. Panelist response area: After assessing the samples in line with the criteria, panelists will require a location to record their remarks. In various testing procedures, panelists may fill in or circle the 3-digit code of the sample(s) they've chosen, or indicate their opinion on a line or scale. The hunt for panelists is still ongoing. Panelists should be picked based on their proximity to the testing location and knowledge of the topic matter to achieve the best results. Customers of a brewery or distilery, as well as people of the local community, maybe frequent users of your crop or product. While there are other methods for recruiting, most operators who wish to test repeatedly will utilize an email listserv system (Peinado et al., 2014). You'll need to contact as many potential panelists as possible so that they are instantly alerted of any paneling opportunities for which they are qualified. Flyers and social media posts can help increase effectiveness. Consent that has been informed Keep in mind that panelists have all of the information they need to make an informed decision about whether or not to participate in a panel discussion.

At all times, researchers must prioritize the well-being of their subjects. In academic settings, the human subjects institutional review board (IRB) supervises this (IRB). Because the protocols in this series are intended for small-to-medium-scale producers to use on horticultural crops and other products that are regularly consumed by the public, the authors recommend that as long as these guidelines are followed, any resulting studies pose no risk to panelists "beyond the ordinary risks of daily life." Screening Participants on the panel should be subjected to a comprehensive vetting procedure before being permitted to speak. Panelists' grasp of the product or feature under consideration is all that is necessary here. When dealing with regular consumers or clients, this is simple, but when dealing with members of the community, it becomes more difficult.

According to a study, the questions in the questionnaires might be used to identify panelists. Question-wording must be carefully crafted to ensure that no exam specifics are divulged. As an alternative to asking if they drink regularly, use a multiple-choice question. If one of their answers included alcohol, they would be qualified to answer that question. To avoid bias in the panelists' replies to the screening questions, any facts that are being screened for should not be included in the exam title. Panelists' incentives an incentive or recompense, such as a monetary payment or a complementary product, is commonly used to encourage panelist participation. In most circumstances, this incentive is necessary, but it must not interfere with testing. Panelists should be able to opt-out of participating if they so want, rather than being forced to do so by their jobs, organizations, or other obligations (Perera et al., 2014). The incentive must be strong enough to entice a sufficient number of individuals to engage, but not so strong that it becomes the major motivation for those who choose to join. Participants, who are not motivated by a financial incentive, such as brand loyalty or a desire to see a better product will be less enthusiastic during testing, therefore look for people who are enthusiastic about participating in panels for other reasons.

10.3. PREPARATION OF THE SPECIMEN

In sample preparation, the most critical step is standardization, which means making sure that all variables apart from the one being studied are the same in every sample. This ensures that panelists utilize the tested feature to distinguish across samples rather than generating conclusions based on other observations. Reusable vs. disposable containers are important considerations when preparing samples for distribution. Glass and plastic wine glasses may be washed and reused several times, but this requires the time and resources needed before and after the test to prepare and clean them. Figure 6 shows a set of wine glasses with plastic lids, for example (Predieri et al., 2021). The plastic coverings contain volatile components that would otherwise be lost when the panelists walk through the samples, and the stemmed glasses are more in line with how wine is typically consumed. There are several drawbacks to using disposable containers, such as the 2 oz. plastic which can be scribbled on immediately. It's up to the operator, but sensory experts often utilize a variety of containers to find the optimum fit for the product they're testing.

Figure 10.10: Sensory organs.

Source: https://media.istockphoto.com/vectors/educational-worksheet-for-kids-game-for-kids-to-compare-and-connect-vector-id1191312578?k=20&m=1191312578&s=612x612&w=0&h=m2R0Bc3AxGOpPfVXPRIOuVys-SpX-c2aPI-PbcbqQKWg=.

Serve ware should be devoid of any smells, odors, or marks that might affect the panelist's judgments, regardless of whether it is reusable or disposable. Containers can be marked with markers, especially with disposable cups, but they must be branded before they are used for paneling so that the marker aroma can disperse. Most sensory experts, on the other hand, utilize labeling guns that look like those used in supermarkets to apply prices. Using these labels is significantly more efficient than writing by hand, and there are no unpleasant scents to worry about if you're using them on re-usable containers. Dimensions and shape of the sample, it is important to standardize the appearance of samples if they are not being evaluated. Samples that are within 10% of the standard size can be selected using a suggested procedure.

Additional considerations include the crop or product being tested, as well as the test goal. Can you tell me about the typical serving size of this dish? Consumed or evaluated? And if eaten in parts, should it be done in one go or broken up into smaller doses? Do individual units of this product have substantial differences or can it be readily made uniform? When it comes to product testing, operators should make every attempt to give panelists a sample that is both "representative" and "composite." Larger items need collecting a representative sample, such as a slice, cube, or another component (O'Mahony, 2018). Panelists like to have a composite sample of numerous tiny objects, such as hemp flower trimmings or hop cone pellets because this lowers the variation between individual units. In the event that you're unsure, offer panelists too much product rather than too little. The serving temperature In contrast to other commodities, horticultural crop samples are often delivered at room temperature.

As a result, the samples must be supplied at the specified temperature, regardless of how they were heated or cooled. For example, panelists will be able to detect the temperature difference between frozen and freshly harvested hop cones if the former are exhibited with the latter at room temperature before they are compared. Using a thermometer before presenting non-room temperature samples can help ensure that they are consistent. Efforts made to serve Aside from sample preparation, serving, and holding times are equally critical and should be consistent.

Time spent removing hemp flower trimmings from storage and evaluating them by panelists should be consistent between samples since this might affect volatile oil concentrations and overall scent intensity. Serving a carbonated beverage should be done consistently; otherwise, the amount of carbonation might fluctuate. Ice cubes should be added to beverages at the same time, in the same amount, to reduce melting fluctuations while serving. Reviving the taste buds Palate cleaning is a sensory assessment technique that involves removing all traces of a sample's previous flavor, aroma, or sensation before moving on to the next. To test oral cleansers, and used a variety of sweet (jelly beans), bitter (coffee), and fatty meals (sausage) Moreover, they discovered that table water crackers were the only cleanser that worked with all of the items they examined (Nishinari & Fang, 2018). Using distiled or deionized water is the best way to eliminate any conflicting tastes from the water and cracker tray. To avoid nasal congestion when sniffing samples, use the back of your hand or forearm.

Figure 10.11: Sensory evaluation.

Source: https://media.istockphoto.com/photos/professional-q-grader-preparing-to-test-and-inspecting-the-quality-of-picture-id1210600057?k=20&m=1210600057&s=612x612&w=0&h=UeOmKiEC37sWohLnQKYcwUEBQ-RT-4DzWPZ7Ype6uyrA=.

10.4. A CATEGORY'S EXTENSIBILITY

In some cases, coded samples are displayed continuously or sequentially in a balanced sequence, depending on the preference of the panel member. Gradation scales are often used to analyze several qualities of food by using a series of words structured in increasing or decreasing intensity (e.g., sweetness, off-flavor, etc.). Following that, each point on the scale is assigned a number, with the lowest number denoting zero intensity. This data can be used to determine whether there are any statistically significant differences in the mean intensity ratings of the various samples, for example, by performing an analysis of variance. In this experiment, we're looking for a link between sensory and physical acuity. Another possible application for this tool is comparing the quality ratings of two or more goods (Nwakalor & Obi, 2014). Participants in this activity can choose from a wide range of numbers, with the characteristic that ratios or proportions between the numerical assignments reflect sensory intensity ratios. In the first sample, subjects can select any value other than zero or a negative integer.

The ratings given to subsequent samples must be proportionate to the ratings given to the first sample. Following normalization, the assigned numbers are statistically analyzed. Investigating the Flavor Profile You'll be able to describe the product's scents and flavors, as well as its sensory attributes and aftertastes, using this method. It is done sequentially, with a constant scale used to assign the strength of each olfactory and gustatory note. Four or five panelists are frequently used. Following that, the moderator of the panel presents the results of the individual panelists to the group for discussion (Muhimbula et al., 2011). The final flavor profile can reflect the impact of these changes on the overall taste after changes in components, processing, packaging, and age.

A study of a material's texture this procedure is described using the flavor profile approach. Mechanical, geometric, lipid, and moisture properties are all considered. The panel will consist of six to nine people. The findings of the panel are recorded and compared to a profile to determine similarities and differences.

Analysis of Descriptive and Quantitative Data This method, which employs an unstructured category scale and a minimum of six trained panelists, necessitates repeated assessments of each test product. An example of a cookie is shown below.

10.5. DESCRIPTIVE TESTS CAN BE USED TO EVALUATE COOKIES

The three main types of affective testing are pair performance tests, ranking tests, and rating scales.

- Performance Evaluation – Paired Evaluations Pair-performance tests for choice are most commonly used when two samples are presented concurrently or sequentially. Panelists may justify their preference decision in addition to their response based on the qualities listed if desired. This method can be used to compare a standard product to a large number of experimental items in a sample series.
- The Ranking Examination – With this new version, the paired-preference test has been improved. Three or more coded samples are displayed at the same time to allow the panelist to double-check their original decision. The respondents are asked to arrange the samples in the order that they prefer (Munasinghe

et al., 2013). This method is unable to determine the degree of favorability or disfavor ability for some samples. Scale ratings on the evaluation scale represent panelists' perceptions of the intensity of a characteristic concerning its context. There are currently several rating scales in use.

- The Hedonic Rating Scale – The purpose of this survey is to determine how much a community values food products. This method can be used to determine whether something is present or accepted. The technique is dependent on respondents' ability to clearly and consistently define their likes and dislikes. Several variations of the nine-point hedonic scale have been successfully applied such as;
  - The number of rating categories is limited to five.
  - Use of a non-numerical, unstructured line scale with "like" and "dislike" at opposite ends.
  - Removal of neutral rating categories from cartoons representing pleasure and displeasure levels (facial hedonic scale), and the panelist will rate each sample, and the scales will be appropriately identified. The directives must not influence the panelists' responses. To see if there is a difference in the degree of similarity between or among samples, hedonic scale assessments are converted to numerical scores, and statistical analysis is performed. A hedonic rating test's findings can be both absolute and relative to the test samples (Mouta et al., 2016). Absolute information is provided by the direction and degree of difference between or among the sample scores, whereas relative information is provided by the direction and degree of difference between or among the sample scores. To investigate the rating data from the hedonic scale, rank-sum analysis, the t-test, and the chi-square test are used.

BIBLIOGRAPHY

1. Abamecha, N., (2020). Research review on formulation and sensory evaluation of complementary foods from cereals and legumes in Ethiopia. *Food Science and Nutrition Technology, 5*(5), 1–7.
2. Abdulmajid, N., Bautista, M. K., Bautista, S., Chavez, E., Dimaano, W., & Barcelon, E., (2014). Heavy metals assessment and sensory evaluation of street vended foods. *International Food Research Journal, 21*(6).
3. Abebe, Y., Stoecker, B. J., Hinds, M. J., & Gates, G. E., (2006). Nutritive value and sensory acceptability of corn-and kocho-based foods supplemented with legumes for infant feeding in Southern Ethiopia. *African Journal of Food, Agriculture, Nutrition and Development, 6*(1), 1–19.
4. Anton, A. A., & Luciano, F. B., (2007). Instrumental texture evaluation of extruded snack foods: A review evaluación instrumental de textura en alimentos extruidos: Una revisión. *CYTA-Journal of Food, 5*(4), 245–251.
5. Ayseli, M. T., Yilmaz, M. T., Cebi, N., Sagdic, O., Ozmen, D., & Capanoglu, E., (2020). Physicochemical, rheological, molecular, thermal and sensory evaluation of newly developed complementary infant (6–24 months old) foods prepared with quinoa (*Chenopodium quinoa* Willd.) flour. *Food Chemistry, 315*, 126208.
6. Baguley, D., McFerran, D., & Hall, D. (2013). Tinnitus. *The Lancet*, *382*(9904), 1600-1607.

7. Beekman, T. L., Luthra, K., Jeesan, S. A., Bowie, R., & Seo, H. S., (2022). Should panelists refrain from wearing a personal fragrance prior to sensory evaluation? The effect of using perfume on olfactory performance. *Foods, 11*(3), 428.
8. Biró, B., Sipos, M. A., Kovács, A., Badak-Kerti, K., Pásztor-Huszár, K., & Gere, A., (2020). Cricket-enriched oat biscuit: Technological analysis and sensory evaluation. *Foods, 9*(11), 1561.
9. Birol, E., Meenakshi, J. V., Oparinde, A., Perez, S., & Tomlins, K., (2015). Developing country consumers' acceptance of biofortified foods: A synthesis. *Food Security, 7*(3), 555–568.
10. Bourgeois, A. E., & Bourgeois, J. O. (1970). Theories of olfaction: A review. *Revista Interamericana de Psicología/Interamerican Journal of Psychology*, *4*(1).
11. Bodyfelt, F. W., Drake, M. A., & Rankin, S. A., (2008). Developments in dairy foods sensory science and education: From student contests to impact on product quality. *International Dairy Journal, 18*(7), 729–734.
12. Briggs, M. H., & Duncan, R. B. (1961). Odour receptors. *Nature*, *191*(4795), 1310-1311.
13. Castro, W. F., Cruz, A. G., Bisinotto, M. S., Guerreiro, L. M. R., Faria, J. A. F., Bolini, H. M. A., & Deliza, R., (2013). Development of probiotic dairy beverages: Rheological properties and application of mathematical models in sensory evaluation. *Journal of Dairy Science, 96*(1), 16–25.
14. Chen, H., Wu, Y., Wang, J., Hong, J., Tian, W., Zhao, D., & Sun, B., (2022). Uncover the flavor code of roasted sesame for sesame flavor baijiu: Advance on the revelation of aroma compounds in sesame flavor baijiu by means of modern separation technology and molecular sensory evaluation. *Foods, 11*(7), 998.
15. Chen, L., & Opara, U. L., (2013). Texture measurement approaches in fresh and processed foods—A review. *Food Research International, 51*(2), 823–835.
16. Chira, K., Anguellu, L., Da Costa, G., Richard, T., Pedrot, E., Jourdes, M., & Teissedre, P. L., (2020). New C-glycosidic ellagitannins formed upon oak wood toasting, identification and sensory evaluation. *Foods, 9*(10), 1477.

17. Chowdhury, T., & Das, M., (2015). Sensory evaluation of aromatic foods packed in developed starch based films using fuzzy logic. *International Journal of Food Studies, 4*(1).
18. Coggins, P. C., (2007). Sensory methodology of muscle foods. *Handbook of Meat, Poultry and Seafood Quality,* 61–70.
19. Daher, D., Deracinois, B., Baniel, A., Wattez, E., Dantin, J., Froidevaux, R., & Flahaut, C., (2020). Principal component analysis from mass spectrometry data combined to a sensory evaluation as a suitable method for assessing bitterness of enzymatic hydrolysates produced from micellar casein proteins. *Foods, 9*(10), 1354.
20. De Kock, H. L., & Magano, N. N., (2020). Sensory tools for the development of gluten-free bakery foods. *Journal of Cereal Science, 94*, 102990.
21. Difonzo, G., Squeo, G., Calasso, M., Pasqualone, A., & Caponio, F., (2019). Physico-chemical, microbiological and sensory evaluation of ready-to-use vegetable pâté added with olive leaf extract. *Foods, 8*(4), 138.
22. Duncan, H. J., Doty, R. L., & Murphy, C. (2003). Clinical disorders of olfaction. In *Handbook of olfaction and gustation* (pp. 822-849). CRC Press.
23. Ebuehi, O. A. T., & Oyewole, A. C., (2007). Effect of cooking and soaking on physical characteristics, nutrient composition and sensory evaluation of indigenous and foreign rice varieties in Nigeria. *African Journal of Biotechnology, 6*(8).
24. Elina, M., Theobald, C. M., Cornelio, N. N., & Henry, L., (2016). Sensory evaluation of extruded quality protein maize-based supplementary foods. *African Journal of Food Science, 10*(7), 105–111.
25. Elsamani, M. O., Habbani, S. S., Babiker, E. E., & Ahmed, I. A. M., (2014). Biochemical, microbial and sensory evaluation of white soft cheese made from cow and lupin milk. *LWT-Food Science and Technology, 59*(1), 553–559.
26. Feng, T., Wu, Y., Zhang, Z., Song, S., Zhuang, H., Xu, Z., & Sun, M., (2019). Purification, identification, and sensory evaluation of kokumi peptides from *Agaricus bisporus* mushroom. *Foods, 8*(2), 43.
27. Finkelstein, S. R., & Fishbach, A. (2010). When healthy food makes you hungry. *Journal of Consumer Research*, *37*(3), 357-367.

28. Fiorentini, M., Kinchla, A. J., & Nolden, A. A., (2020). Role of sensory evaluation in consumer acceptance of plant-based meat analogs and meat extenders: A scoping review. *Foods, 9*(9), 1334.

29. Gunaratne, T. M., Fuentes, S., Gunaratne, N. M., Torrico, D. D., Gonzalez, V. C., & Dunshea, F. R., (2019). Physiological responses to basic tastes for sensory evaluation of chocolate using biometric techniques. *Foods, 8*(7), 243.

30. Happiness, S. M., Abdulsudi, I. Z., & Joyce, K., (2011). Formulation and sensory evaluation of complementary foods from local, cheap and readily available cereals and legumes in Iringa, Tanzania. *African Journal of Food Science, 5*(1), 26–31.

31. Haagen-Smit, A. J. (1952). Chemistry and physiology of Los Angeles smog. *Industrial & Engineering Chemistry*, *44*(6), 1342-1346.

32. Mothersbaugh, D. L., Hawkins, D. I., Kleiser, S. B., Mothersbaugh, L. L., & Watson, C. F. (2020). *Consumer behavior: Building marketing strategy*. New York, NY, USA: McGraw-Hill Education. 11th Ed.

33. Hong, S. P., Kim, Y. H., Yang, J. N., Chae, I. S., & Shin, D. B., (2008). Sensory evaluation of dining staffs at UCLA for Korean foods. *Journal of the Korean Society of Food Culture, 23*(6), 705–712.

34. Huang, Y., Pu, D., Hao, Z., Liang, L., Zhao, J., Tang, Y., & Zhang, Y., (2022). Characterization of taste compounds and sensory evaluation of soup cooked with sheep tail fat and prickly ash. *Foods, 11*(7), 896.

35. Irmak, C., Vallen, B., & Robinson, S. R. (2011). The impact of product name on dieters' and nondieters' food evaluations and consumption. *Journal of consumer research*, *38*(2), 390-405.

36. Igarashi, A., Kawasaki, M., Nomura, S. I., Sakai, Y., Ueno, M., Ashida, I., & Miyaoka, Y., (2010). Sensory and motor responses of normal young adults during swallowing of foods with different properties and volumes. *Dysphagia, 25*(3), 198–206.

37. Jain, M., Singh, C., Gupta, K., & Jain, P., (2014). Formulation and sensory evaluation of healthy heart foods. *Nutraceuticals, 3*, 1–5.

38. Kang, S. W., Choi, G. H., Yi, S. Y., Kim, S. S., Song, B. S., Kim, J. H., & Le, J. W., (2009). Analysis of the sensory evaluation for the Korean space foods performed in ISS. *Journal of the Korean Society for Aeronautical & Space Sciences, 37*(8), 819–827.

39. Kohyama, K., (2020). Food texture–sensory evaluation and instrumental measurement. *Textural Characteristics of World Foods,* 1–13.

40. Koli, D. K., Rudra, S. G., Bhowmik, A., & Pabbi, S., (2022). Nutritional, functional, textural and sensory evaluation of spirulina enriched green pasta: A potential dietary and health supplement. *Foods, 11*(7), 979.
41. Kukovics, S., & Németh, T., (2014). Sensory evaluation of various lamb meat foods by judging consumers. *Acta Alimentaria, 43*(2), 254–263.
42. Kweku, A. F., Mutukumira, A. N., Brough, L., Weber, J. L., Hardacre, A., & Coad, J., (2013). Carbohydrate composition, viscosity, solubility, and sensory acceptance of sweet potato-and maize-based complementary foods. *Food & Nutrition Research, 57*(1), 18717.
43. Lee, H. S., Song, M. W., Kim, K. T., Hong, W. S., & Paik, H. D., (2021). Antioxidant effect and sensory evaluation of yogurt supplemented with hydroponic ginseng root extract. *Foods, 10*(3), 639.
44. Lee, J., (2013). Globalization of Korean foods and sensory evaluation. *Food Science and Industry, 46*(3), 29–39.
45. Lomarat, P., Moongkarndi, P., Jaengprajak, J., Samer, J., & Srisukh, V., (2019). Three functional foods from *Garcinia mangostana* L. using low-alpha-mangostin aqueous extract of the pericarp: Product development, bioactive compound extractions and analyses, and sensory evaluation. *Thai Journal of Pharmaceutical Sciences (TJPS), 43*(1).
46. López-López, A., Sánchez-Gómez, A. H., Montaño, A., Cortés-Delgado, A., & Garrido-Fernández, A., (2019). Panel and panelist performance in the sensory evaluation of black ripe olives from Spanish manzanilla and hojiblanca cultivars. *Foods, 8*(11), 562.
47. Lotfy, T. M., El-Hendy, H. A., & Shawir, S., (2021). Effect of mushroom, oat and their mixtures as functional foods on antioxidant activities, hematological parameters of obese rats and its sensory evaluation. *Alexandria Science Exchange Journal, 42*, 21–36.
48. Milner, T., & Rosenstreich, D. (2013). A review of consumer decision-making models and development of a new model for financial services. *Journal of Financial Services Marketing, 18*(2), 106-120.
49. Mouta, J. S., De Sá, N. C., Menezes, E., & Melo, L., (2016). Effect of institutional sensory test location and consumer attitudes on acceptance of foods and beverages having different levels of processing. *Food Quality and Preference, 48*, 262–267.

50. Muhimbula, H. S., Issa-Zacharia, A., & Kinabo, J., (2011). Formulation and sensory evaluation of complementary foods from local, cheap and readily available cereals and legumes in Iringa, Tanzania. *African Journal of Food Science, 5*(1), 26–31.
51. Munasinghe, M. A. D. D., Silva, K. F. S. T., Rasika, D. M. D., Jayarathne, M. P. K., & Sarananda, K. H., (2013). Formulation and sensory evaluation of yoghurt-based weaning foods manufactured from mung bean, soybean and brown rice. *International Journal of Scientific and Research Publications, 3*(4), 1–9.
52. Nishinari, K., & Fang, Y., (2018). Perception and measurement of food texture: Solid foods. *Journal of Texture Studies, 49*(2), 160–201.
53. Nwakalor, C. N., & Obi, C. D., (2014). Formulation and sensory evaluation of sorghum based weaning food fortified with soybean and unripe plantain flour. *International Journal of Nutrition and Food Sciences, 3*(5), 387–390.
54. O'Mahony, M., (2018). Sensory difference and preference testing: The use of signal detection measures. In: *Applied Sensory Analysis of Foods* (pp. 145–175). Routledge.
55. Peinado, I., Girón, J., Koutsidis, G., & Ames, J. M., (2014). Chemical composition, antioxidant activity and sensory evaluation of five different species of brown edible seaweeds. *Food Research International, 66*, 36–44.
56. Perera, M. P. M. S. H., Sivakanesan, R., Abeysekara, D. T., & Sarananda, K. H., (2014). Sensory evaluation, proximate analysis and available carbohydrate content of soy flour incorporated cereal based traditional Sri Lankan breakfast foods. *Int. J. Res. Agric. Food Sci., 1*, 2311–2476.
57. Predieri, S., Magli, M., Gatti, E., Camilli, F., Vignolini, P., & Romani, A., (2021). Chemical composition and sensory evaluation of saffron. *Foods, 10*, 2604.
58. Pu, D., Zhang, Y., Zhang, H., Sun, B., Ren, F., Chen, H., & Tang, Y., (2020). Characterization of the key aroma compounds in traditional hunan smoke-cured pork leg (Larou, THSL) by aroma extract dilution analysis (AEDA), odor activity value (OAV), and sensory evaluation experiments. *Foods, 9*(4), 413.
59. Rousta, N., Hellwig, C., Wainaina, S., Lukitawesa, L., Agnihotri, S., Rousta, K., & Taherzadeh, M. J., (2021). Filamentous fungus *Aspergillus oryzae* for food: From submerged cultivation to fungal

burgers and their sensory evaluation—A pilot study. *Foods, 10*(11), 2774.

60. Rowe, M., Young, L. S., Crocker, J., Stokes, H., Henderson, S., & Rickinson, A. B. (1991). Epstein-Barr virus (EBV)-associated lymphoproliferative disease in the SCID mouse model: implications for the pathogenesis of EBV-positive lymphomas in man. *The Journal of experimental medicine, 173*(1), 147-158.
61. Schilling, M. W., & Pham, A. J., (2012). Sensory evaluation of muscle foods. In: Hui, Y. H., (ed.), *Handbook of Meat and Meat Processing* (2nd edn., pp. 207–224).
62. Serventi, L., Brennan, C., & Mustafa, R., (2022). Physicochemical and sensory evaluation of grain-based food. *Foods, 11*(9), 1237.
63. Shinde, K. J., Pardeshi, I. L., Pawar, S. G., & Gajakos, A. V., (2016). Applications of fuzzy logic technique in sensory evaluation of ready to eat foods. *Journal of Ready to Eat Food, 3*(04), 45–50.
64. Short, E. C., Kinchla, A. J., & Nolden, A. A., (2021). Plant-based cheeses: A systematic review of sensory evaluation studies and strategies to increase consumer acceptance. *Foods, 10*(4), 725.
65. Silva, F., Duarte, A. M., Mendes, S., Magalhães, E., Pinto, F. R., Barroso, S., & Gil, M. M., (2020). Seasonal sensory evaluation of low commercial value or unexploited fish species from the Portuguese coast. *Foods, 9*(12), 1880.
66. Sun, Y., Ma, Y., Chen, S., Xu, Y., & Tang, K., (2021). Exploring the mystery of the sweetness of baijiu by sensory evaluation, compositional analysis and multivariate data analysis. *Foods, 10*(11), 2843.
67. Sutrisna, A., Vossenaar, M., Izwardy, D., & Tumilowicz, A., (2017). Sensory evaluation of foods with added micronutrient powder (MNP) "*Taburia*" to assess acceptability among children aged 6–24 months and their caregivers in Indonesia. *Nutrients, 9*(9), 979.
68. Tsai, Y. J., Lin, L. Y., Yang, K. M., Chiang, Y. C., Chen, M. H., & Chiang, P. Y., (2021). Effects of roasting sweet potato (*Ipomoea batatas* L. Lam.): Quality, volatile compound composition, and sensory evaluation. *Foods, 10*(11), 2602.
69. Tuorila, H., & Monteleone, E., (2009). Sensory food science in the changing society: Opportunities, needs, and challenges. *Trends in Food Science & Technology, 20*(2), 54–62.

70. Van Ittersum, K., & Wansink, B. (2012). Plate size and color suggestibility: The Delboeuf illusion's bias on serving and eating behavior. *Journal of Consumer Research*, *39*(2), 215-228.
71. Warnakulasooriya, K. M. H. K., Bandara, J. D. S. A. C. S., & Sudantha, B. H., (2020). Sensory evaluation of foods using modern AI techniques and brain wave analysis. In: *2020 International Conference on Image Processing and Robotics (ICIP)* (pp. 1–4). IEEE.
72. Yağcı, S., & Göğüş, F., (2008). Response surface methodology for evaluation of physical and functional properties of extruded snack foods developed from food-by-products. *Journal of Food Engineering*, *86*(1), 122–132.
73. Yang, X., & Boyle, R. A., (2016). Sensory evaluation of oils/fats and oil/fat–based foods. In: *Oxidative Stability and Shelf Life of Foods Containing Oils and Fats* (pp. 157–185). AOCS Press.

INDEX

A

Acetaldehyde 186
additives 131, 138, 144
Affective testing 157
agribusinesses 180
Analytical sensory evaluation 157
Analytical testing 157
Anatomy 111, 119
appearance 26, 32, 33, 40, 44
aroma 182, 183, 184, 186, 188
auditory 130

B

bananas 187, 188
Beverage acidity 182
biosciences 156
Blood-borne toxins 111
brand loyalty 4, 8, 12

C

cane sugar 186
cellular interaction 108
Center for Food Safety & Applied Nutrition (CFSAN) 57
chemical analysis 27
Chemical-physical measurement 218
chemoreceptors 80, 87, 101
chemosensory signals 109
cider 185, 186
cilia 111, 114
coffee 230, 231, 237
Communication 109
competitive advantage 131, 139
consistency 26, 40, 44, 46
consumer acceptability 52, 53, 55, 58, 59, 64, 65, 66, 67, 68, 69, 72, 73, 78
Consumer acceptability 53, 65
consumer behavior 58
Consumer sensory panelists 206
corporate social responsibility 58, 59
Corporate sustainability 57, 58
customer perception 180
Customer safety 26
customer segmentation 3

D

dating 108
decision-making 2, 3, 4, 5, 6, 7, 8, 9, 10, 12, 13, 17, 18, 23

decision-making process 4, 5, 6, 7, 9, 12, 17, 23
decision-making styles 2
Descriptive sensory evaluation 158
Diachasmimorpha longicaudata 109
dietary 26, 32, 38
discernment 80, 85, 90, 92, 95, 97
discrimination testing 163
Discrimination tests 164
distinctive automatic behaviors 83, 84
distribution channels 52, 65, 67
Dlg-based sensory assessment 209
Dual Standard Test 168
duo-trio test 164, 167, 170

E

Electrical signals 83
electronic device technology 157
environmental friendliness 27, 35
Environmental Protection Agency (EPA) 16
Environmental responsibility 59
Ester molecules 183
Eyesight 83

F

feeding 108, 118
flavor 26, 32, 33, 40, 43, 46, 130, 131, 132, 133, 134, 136, 137, 138, 142, 144, 146, 148, 149, 151
Food and Drug Administration (FDA) 16, 24
Food and Drug Administration (FDA) regulations 56
food packaging materials 28
Food quality 26, 27, 32, 33, 34, 36, 39, 42
food quality programs 205
food safety 26, 27, 28, 30, 32, 37, 38, 44
Food sensory analysis 131
food system 131, 135
fungicide fertilizers 28

G

gas chromatography 131, 135, 220
genetically modified organisms 28, 38
glands 111
goblet cells 111
grading system 204
grapefruits 187
groceries 2

H

Hedonic Rating Scale 240
hedonic sensory testing 209
high-performance liquid chromatography (HPLC) 220
human senses 80, 81, 82, 83, 103
human sensory organs 83
hygiene standards 26

I

ingredients 26, 33, 35, 44
inhomogeneous meals 222
Innovation 55, 69, 70
institutional review board (IRB) 234
integrated crop management (icm) 28
intense competition 11
International Food Information Council (IFIC) 14, 15
isoamyl acetate 183

L

laundry detergent 10
limited decision-making 4, 11

M

marketing 3, 9, 11
marketing strategies 3, 11
market research 52, 77
Meal quality 26
meat samples 204
mechanoreceptors 80, 87, 96
Molecular Organization of the Brain (MOB) 108
Mucous 111
multiple sample test 164
multiple standard tests 164

N

nasal mucosa 111
neuropsychiatric depression 108
newspapers 6
nitrogen fertilizers 28
null hypothesis 163
nutritional information 27
nutrition education campaigns 3

O

olfactory 130
olfactory circuit 108
olfactory epithelium 111, 112, 114, 115, 116, 120, 121, 126, 127, 128
olfactory receptor neurons 108
Olfactory sensory neuroepithelial cells 108
olfactory sustentacular cells 108
olfactory system 108, 113, 114, 118, 120
organic wines 180, 181, 197, 198, 199, 201
oxygen 187, 188, 192, 193

P

paired comparison test 164, 165, 167, 170
parasitoid wasps 109
parenting 108
pears 187
Performance Evaluation 239
personality 2
pesticides 28
photoreceptors 80, 87, 90
plywood 232
Poor sanitation 27
product development 205, 209, 210
product positioning 52
product texture 134
professional sensory laboratory 221
progenitor cells 108
Psychological challenges 53
psychology 2

Q

Quality assurance 55
quality control 130, 131, 134, 135, 140, 144
quality management system (qms) 28
Quantitative Descriptive Analysis (QDA) 159

R

Ranking Examination 239
raspberries 187
rating system 162
raw materials 131, 138, 144

Regulatory compliance 56

S

Salinity 182
Sample preparation 131
Securities and Exchange Commission (SEC) 56
Sensation 80
sense 80, 81, 83, 84, 85, 86, 87, 90, 91, 92, 94, 95, 96, 97, 98, 99, 100, 101, 102, 104, 105
sensory abnormalities 205
Sensory analysis 131, 134, 136, 137, 139, 140, 141, 144, 152, 153
Sensory assessment 134, 135, 139, 153
sensory aversions 220
sensory evaluation 130, 131, 132, 133, 134, 135, 136, 138, 139, 144, 154
Sensory experts 156
Sensory hedonic assessment 209
sensory organ 80
Sensory panelists 206
sensory panels 205, 206, 214
sensory preferences 220
Sensory quality control 132
sensory sensitivity 220, 229
Sensory systems 82
shopping personality 2
sickness 83
Sight 83, 104
skin 82, 83, 86, 95, 97, 101, 102, 103
smell 130, 133, 135, 136, 137, 138, 143, 147, 148, 151, 153
social media accounts 13
sound or 80
standard sensory practices 218
starvation 83, 84
Statistical mathematics 134
statistics 156, 176
storage conditions 131, 135, 145
strawberries 187
suffocation 83, 84, 101
sustainable brand 58
sustainable food production 14

T

tactilely 130
tangible receptor cells 80
testing 157, 160, 161, 162, 163, 164, 166, 167, 172, 173, 174
texture 26, 40, 42, 43, 44, 46
thermoreceptors 80, 87, 96, 97
thirst 83, 84, 101
tongue 83, 95, 98, 99
Touch 83, 104
triangle test 164, 167, 169
trifold poster boards 232

V

vascular tissue 111
vidian nerve 111
visual appearance 134
vocabulary 159

W

Water 185

Y

yeast 181, 183, 184, 185, 186, 187, 190, 192, 194, 195, 196

Z

zebrafish 109